Water Quality In A Stressed Environment

Water Quality

Readings In Environmental Hydrology

In A Stressed Environment

Edited by
Wayne A. Pettyjohn
The Ohio State University
Columbus, Ohio

BURGESS PUBLISHING COMPANY
MINNEAPOLIS, MINNESOTA

Dedicated to my wife Sue
and Morris "Moe" Deutsch

Photo on title page and on page 224 courtesy of *Minneapolis Tribune*, Minneapolis, Minn.

Preface

Water pollution has received wide coverage in news media, magazines, and books in the past several years, but it is difficult to obtain reports that either adequately describe specific examples of pollution or provide background geologic or hydrologic information. These readings are intended to fill this need. This book can be used as supplementary reading material for students in geology, hydrology, engineering, and perhaps social studies and law, as well as in newly emerging environmental courses.

The papers have been selected from several sources and, although their implications are wide, they deal mainly with ground-water and surface-water contamination. Some topics, including drainage of acid water from coal-mining areas, on which much of the literature is highly technical, are only briefly described. Other subjects, such as thermal pollution, are omitted entirely because adequate coverage for the lay reader is readily available elsewhere. In addition, a great deal of information relating to contamination of surface waters by municipal and industrial wastes has received wide coverage for several years and, consequently, this subject is only briefly treated here. It is the aim of this source book not to provide a complete set of case histories but to point out only a few significant occurrences of water pollution that have been reported over the last 25 years.

These readings have been drawn mainly from scientific journals and governmental publications. Most of the reports were adopted in their entirety, but a few have been edited. Several articles by H.E. LeGrand, Morris Deutsch, and Graham Walton, a few of which are included herein, reflect their long interest in water pollution; other writers have only lately made their impact. Most reports describing the possible adverse effects of trace elements on health also have appeared only recently, mainly since the discovery of mercury in Lake Saint Clair and Lake Erie, but investigations by scientists such as H.V. Warren and his colleagues at the University of British Columbia have been in progress for more than 20 years. Warren's reports, of which one is included here, are both readable and disturbing. They are highly recommended.

Although the various doctrines of water law have been developed by court decisions and statutes over the past several centuries, the concepts are not generally understood by the layman. The new field of environmental law is just beginning to appear, but it has a firm foundation through long consideration by the courts. The three articles on water law included herein have wide application in environmental problems.

It is sincerely hoped that this series of readings will help to dispel several environmental myths, provide a clearer insight into the problems of environmental hydrology, and offer suggestions or techniques that can be used in future water pollution studies.

I wish to express my sincere appreciation to many authors and publishers who have permitted inclusion of their material. I am also indebted to the many individuals whose efforts have influenced my views and selections, even though their own works have not been included. The support received from the staff of the Department of Geology at The Ohio State University is gratefully acknowledged. Special thanks are due S.E. Norris, Water Resources Division, U.S. Geological Survey, for his help and guidance.

August, 1972   Wayne A. Pettyjohn

Contents

Part One

The Water
We Drink

People in the United States have been long blessed with drinking water of good quality, both biologically and chemically. The development of modern water-treatment techniques has reduced the occurrence of waterborne diseases to a very low level. Nonetheless, even though we worry little about the possibility of epidemics of cholera or typhoid fever, nearly every year municipalities have problems with water taste and odor and occasionally are beset with outbreaks of gastrointestinal problems that might be traceable to the water supply.

Investigations in recent years have shown that certain heavy metals or trace elements occur in water supplies, some of which may be toxic. In most instances, however, the concentration in water is very low. On the other hand, very little is known of the long-term health effects resulting from ingestion of minute concentrations of trace elements over a long period. Unfortunately, many of these substances are not removed from water by normal water-treatment processes.

The report by J.E. McDermott brings to light several interesting facts dealing with public water supplies which tend to show that certain treatment techniques, or in some instances inadequately trained operators, do not provide drinking water of acceptable quality. It is disturbing to find that 41 percent of

Photo opposite courtesy of *Minneapolis Tribune,* Minneapolis, Minn.

the 969 systems investigated were delivering water of inferior quality to 2.5 million people.

On the other hand, what are "permissible criteria" or "desirable criteria," and what are the recommended limits of certain chemical and biological constituents and of physical properties in raw drinking water supplies? Answers to questions such as these are found in the report "Water Quality Criteria—Public Water Supplies." The origin and significance of selected water quality parameters are given in the final report in this section by C.N. Durfor and Edith Becker. The latter two reports provide background water-quality information that will aid in understanding the significance of contaminants reported elsewhere in this book.

Significance of the National Community Water Supply Study

James E. McDermott*

Preface

Contemporary American society recognizes a host of interrelated factors that determine the quality of urban life. In addition to the basic needs—food, clothing and shelter—we have recently begun to recognize two other daily necessities that were heretofore thought to be of unquestionable quality and available in unlimited quantities; ample quantities of clean air, from moment to moment, and safe drinking water, from hour to hour.

The Community Water Supply Study concerns the current and future healthfulness and dependability of the drinking water supplied to over 150 million Americans by community water supply systems. The remaining population drinks from private supplies. The purpose of the study was to determine the quality of drinking water being delivered to the over 18 million people in the study areas and the health risk factors that enabled scientists and engineers to evaluate the ability of these systems to continue to provide adequate supplies of safe water now and in the future. The Analysis of National Survey Findings of the National Community Water Supply Study (July 1970) is based on a survey of 969 representative public water supply systems located in nine areas of the Nation. This statement attempts to place the technical findings into a national perspective. It seeks to answer two questions about the nation's water supplies: (1) Are well established standards of good practice being applied to assure the quality and dependability of water being delivered to consumers' faucets today? and, (2) What needs to be done to assure adequate quantities of safe drinking water in the future on a National scale? While our study has helped provide answers to these important questions, not all the discussion that follows in this statement is derived solely from the results of this single investigation.

Reprinted from a report by the Public Health Service, 1970, with permission from the author and the Water Supply Programs Division, Environmental Protection Agency.
*Director, Water Supply Programs Division, Environmental Protection Agency, Washington, D.C.

Background

Americans generally assume that the water from their faucets is healthful, and free of bacterial or chemical contaminants that can bring disease. Usually, the assumption is correct. The drinking water supplies in cities and towns of the United States rank in quality, on the average, among the best in the world. Nevertheless, there is cause for serious concern about our drinking water. There are two good reasons for this paradox.

To begin with, it cannot be maintained that *all* of our drinking water is safe. It is true that the classical communicable waterborne diseases of years past—typhoid fever, amoebic dysentery and bacillary dysentery—were brought under control by the 1930s. However, we still have outbreaks of communicable disease from sewage contamination of water supply systems in the United States. Recent outbreaks are discussed later in this report. As we shall see in this report, we found evidence of bacterially contaminated water being served to consumers in communities ranging in size from less than 500 to 100,000 persons.

Disturbing as it is to find such evidence, there is a second, more far reaching problem of considerable importance to the country. That problem is the ability of all our present municipal water supply systems to continue to deliver water of good quality and adequate quantity in the decades ahead to a rapidly rising population. This is made all the more difficult by the growing amount of chemical pollutants entering our lakes, streams and aquifers.

Current forecasts provide an indication of how much water we will be needing in the future. According to one calculation, we used 270 billion gallons of water per day in 1965 in support of industry, agriculture, and for domestic drinking purposes. By the year 2020, our water requirements are expected to exceed 1300 billion gallons each day. But hydrologists estimate that the total usable surface water supply from rainfall is only 700 billion gallons per day. Even today, when we return our used waters to streams or lakes we find ourselves using them over and over again. The need for multiple reuse of water will become greatly amplified in major sections of the country in years ahead. If the future population growth rate is only half of current projections, and even where desalinization of salt and brackish waters is a practical and economically feasible alternative, major sections of the country will find it increasingly necessary to practice multiple reuse in the years ahead. Much of the future problem relates to the need for having this water available when and where it is needed. For this reason, ground water has emerged as a significant source now accounting for more than 20 percent of the Nation's water supply requirements.

Where both surface and ground sources are insufficient, it will become necessary to directly recycle our wastewaters. This means taking wastewaters and using them over again in a closed system without first discharging them into our streams and lakes. With our present technology we cannot use water in this fashion for drinking, recreation or other intimate uses. It is true that during the past decade, much has been learned about the treatment of wastewaters for removal of some organic substances and bacteria, and processes for renovating wastewaters for direct reuse have even proceeded to the pilot plant stage. But the reuse of wastewaters over and over again presents us with new problems; with present treatment processes, chemicals would be concentrated, and therefore, new treatment processes must be developed; fail-safe warning systems must be found; and new methods must be developed to detect and remove such impurities as the pesticides and viruses which currently are present in almost undetectable concentrations. Little is known about the concentrations of carcinogens, antibiotics or hormones present in wastewaters.

Even though wastewater control efforts will be expanded in the future and are sorely needed to minimize future pollution of our drinking water sources, it is clear that water pollution control efforts alone cannot assure a safe drinking water quality. It is highly unlikely that even the best conventional waste treatment will produce a discharge of drinking water quality. As such, treatment does not remove all of today's known potential toxicants or biological agents prior to discharge. In addition, there are pollutants which have an effect on source of drinking water which are not subject to waste treatment. Such pollutants are found in uncontrolled runoff from our fields and forests, and from chemicals spilled in transportation accidents. Both of these examples adversely affect quality at the community water treatment plant intake. Both today and in the future, delivery of adequate supplies of safe water at the consumer's tap will be dependent upon properly designed, constructed and operated municipal water treatment plants and distribution systems.

Scope of the Study

The National Community Water Supply Study was designed to cover a variety of natural and demographic situations across the country. It surveyed 969 public water systems—in the State of Vermont and in eight standard metropolitan statistical areas—New York, New York; Charleston, West Virginia; Charleston, South Carolina; Cincinnati, Ohio; Kansas City, Missouri-Kansas; New Orleans, Louisiana; Pueblo, Colorado; and San Bernardino-Riverside-Ontario, California. The survey investigated every public water system in each of the designated areas. Twenty-two big city systems in the study areas served over 13 million people. The remaining 947 systems served 5 million people in communities of less than 100,000 people and 760 of those 947 systems each served populations of less than 5,000 people.

The survey was not expected to provide a perfect random sample of water supply systems throughout the country, but the results are reasonably representative of the status of the water supply industry in the United States. As detailed in the Analysis of National Survey Findings, and in the nine supportive reports presenting findings for the specific study areas, the Public Health Service Drinking Water Standards of 1962 were used to evaluate both the current quality of drinking water and the health risks associated with the systems delivering that water.

Each water supply system was investigated to determine the quality of water being delivered to the consumer's tap, the adequacy of physical facilities and operating procedures, and the status of surveillance programs so necessary to the delivery of adequate quantities of safe water on a continuing basis consistent with the U.S. Public Health Service Drinking Water Standards. Two or more water samples, depending on the size of the community population, were analyzed for chemical, bacteriological and other constituents. Each sample indicated the quality of water at a particular point in time, and when all samples from a given system were evaluated together, the average quality of water being served during the study was determined.

The evaluation of each system was designed to identify deficiencies which could lead to a system failure in the future that, in turn, could lead to the delivery of potentially hazardous water quality to the consumer. Past records were studied to determine operational practices, including the frequency of past failures of equipment. The current condition of physical facilities was examined for such deficiencies as inadequate disinfection equipment in the event of an emergency, or finished water reservoirs poorly protected from contamination. The surveillance programs were reviewed with an eye on such problems as collection of bacteriological samples on a regular basis and the regular inspection of the distribution systems to prevent recontamination of the drinking water between the treatment plant and the consumer's tap.

Findings in the Study Areas

Drinking water quality defects and health risk problems involving poor operating procedures, inadequate physical facilities, and poor surveillance activities were found in both large cities and small towns irrespective of geographical location. In general, the larger systems, those serving in excess of 100,000 persons including the 10.4 million people in the cities of New York, Cincinnati, Kansas City, and New Orleans, were delivering an "average" acceptable water quality consistent with the Drinking Water Standards. On this *average basis*, 86 percent of the approximately 18 million people covered by this study, or about 15.5 million served by 59 percent of the 969 systems investigated, were receiving good water during the study. The larger systems also evidenced better operation of treatment and distribution facilities. While sanitary defects were found in larger systems, the overall health risk was generally judged to be low, even though improvements in operational procedures and physical facilities are believed warranted in many instances.

Conversely, 41 percent of the 969 systems were delivering waters of inferior quality to 2.5 million people. In fact, 360,000 persons in the study population were being served waters of a potentially dangerous quality. This was particularly true of community systems serving less than 100,000 persons. Even where average quality was good, occasional samples were found to contain fecal bacteria, lead, copper, iron, manganese and nitrate and a few even exceeded the arsenic, chromium, and selenium limits. After all, people do not drink "average" water. They drink "samples" of water from their kitchen faucets or a drinking fountain at work or play. It is particularly important to note that communities of less than 100,000 people evidenced a prevalence of the water quality deficiencies and health risk potential. Some of the very small communities were even drinking water on a day-to-day basis that exceeded one or more of the dangerous chemical limits, such as selenium, arsenic or lead.

The major findings from the study, in the light of today's water treatment technology are as follows:

Quality of Water Being Delivered

- 36 percent of 2,600 individual tap water samples contained one or more bacteriological or chemical constituents exceeding the·limits in the Public Health Service Drinking Water Standards.
- 9 percent of these samples contained bacterial contamination at the consumer's tap evidencing potentially dangerous quality.
- 30 percent of these samples exceeded at least one of the chemical limits indicating waters of inferior quality.
- 11 percent of the samples drawn from 94 systems using surface waters as a source of supply exceeded the recommended organic chemical limit of 200 parts per billion.

Status of Physical Facilities

- 56 percent of the systems evidenced physical deficiencies including poorly protected groundwater sources, inadequate disinfection capacity, inadequate clarification capacity, and/or inadequate system pressure.
- In the eight metropolitan areas studied, the arrangements for providing water service were archaic and inefficient. While a majority of the population was served by one or a few large systems, each metropolitan area also contained small inefficient systems.

Operators' Qualifications

- 77 percent of the plant operators were inadequately trained in fundamental water microbiology; and 46 percent were deficient in chemistry relating to their plant operation.

Status of Community Programs

- The vast majority of systems were unprotected by cross-connection control programs, plumbing inspection programs on new construction, or continuing surveillance programs.

Status of State Inspection and Technical Assistance Programs

- 79 percent of the systems were not inspected by State or County authorities in 1968, the last full calendar year prior to the study. In 50 percent of the cases, plant officials did not remember when, if ever, a state or local health department had last surveyed the supply.
- An insufficient number of bacteriological samples were analyzed for 85

percent of the water systems—and 69 percent of the systems did not even
analyze half of the numbers required by the PHS Drinking Water
Standards.

National Significance of the Study Findings

Well established standards of good practice, in terms of the full application of
existing technology, are not being uniformly practiced today to assure good
quality drinking water. While most professionals hold the USPHS Drinking
Water Standards in high esteem, the study shows that an unexpectedly high
number of supplies, particularly those serving fewer than 100,000 people,
exceeded either the mandatory or recommended constituent levels of bacterial
or chemical content, and a surprisingly larger number of systems evidence
deficiencies in facilities, operation and surveillance.

The National significance can be placed in perspective by considering the
size-distribution of municipal water supply systems that were the subject of
comprehensive facilities census conducted during 1963. At that time, 150
million Americans were being served by 19,236 public water supply systems
including 73 million people dependent upon 18,837 small systems, each serving
communities of less than 100,000 people. When these statistics are compared
with the fact that over 40 percent of the small systems investigated during the
current study evidenced current quality deficiencies *on the average* and both
large and small communities were judged to be giving inadequate attention to
quality control factors, there can be little doubt that this situation warrants
major National concern.

Most of our municipal water supply systems were constructed over 20 years
ago. Since they were built, the populations that many of them serve have
increased rapidly—thus placing a greater and greater strain on plant and
distribution system capacity. Many systems are already plagued by an insufficient supply, inadequate transmission or pumping capacity, and other known
deficiencies that become most evident during peak water demand periods.
Moveover, when these systems were built, not enough was known to design a
facility for the removal of toxic chemical or virus contaminants. They were
designed solely to treat raw water of high quality for the removal of coliform
bacteria. Such facilities are rapidly becoming obsolete as demands rise for water.
The task in the future for our water treatment plants can be visualized by
examining our population trend. By the year 2000—only 30 years from
now—our present population of about 205 million is expected to spurt to 300
million. By that time, it is expected that 187 million people (the total U.S.
population just eight years ago) will be concentrated in four urban agglomera-
tions—on the Atlantic Coast, the Pacific Coast, on the coast of the Gulf of
Mexico and on the shores of the Great Lakes. Most of the remaining population
will be living in cities of 100,000 or more.

In the past, communities and industries were in the favorable position of
being able to select the best source of supply consistent with their quantity and

quality requirements. The demand for more water to quench the thirst of a growing population and meet the needs of expanding industry have led many people to ask how future quantity requirements will be satisfied. Concurrently, expanding water use comes at a time of greatly increased pollution of ground water aquifers, as well as streams, lakes and rivers. Historically and traditionally, ground water coming from its natural environment has been considered of good sanitary quality—safe to drink, if palatable. Nevertheless, 9 percent of the wells sampled during this survey showed coliform bacterial contamination. It seems fair to say that a similar situation prevails nationwide.

Chemical contaminants in our environment have been on the increase for about 25 years, due to the dramatic expansion in the use of chemical compounds for agricultural, industrial, institutional and domestic purposes. There are about 12,000 different toxic chemical compounds in industrial use today, and more than 500 new chemicals are developed each year. Wastes from these chemicals—synthetics, adhesives, surface coatings, solvents and pesticides—already are entering our ground and surface waters, and this trend will increase. We know very little about the environmental and health impacts of these chemicals. For example, we know very little about possible genetic effects. We have difficulty in sampling and analyzing them—we have much greater difficulties in determining their contribution to the total permissible body burden from all environmental insults.

Consideration of the findings of this study leaves no doubt that many systems are delivering drinking water of marginal quality on the average, and many are delivering poor quality in one or more areas of their water distribution systems today. To add to this quality problem, the deficiencies identified with most water systems justifies real concern over the ability of most systems to deliver adequate quantities of safe water in the future.

Recommendations

Modern facilities operated by qualified personnel under adequate surveillance will provide high quality water with the lowest possible risk that current technology can offer. The following recommendations are made to those state and municipal officials concerned with the responsibility for safe, adequate water supply:

- Apply available water treatment and distribution technology, more intensively.
- Determine manpower needs of the state and county programs *now* in order to develop a program to provide technical assistance, training, and adequate surveillance to the Nation's numerous community water supply systems.
- Upgrade the skills of personnel responsible for the operations and maintenance of the water supply systems themselves, particularly in the case of those systems serving fewer than 100,000 people, through short

courses, seminars, and correspondence courses to employees presently
employed in the field as well as those wishing to enter it.

- Expand state laboratory resources to add the capability of routinely
  analyzing water samples for biological and chemical agents of health
  significance.
- Provide educational opportunities in water hygiene at the university level
  to assure the availability of qualified personnel to meet existing and future
  needs.

In addition to defining the need for improvements at the state and
community level, this study's findings also show a need for research, development and planning to improve current practices and to provide adequate supplies
of safe water in the future. The study clearly evidences the need to develop:

- Improved systems including surveillance procedures, to assure continuous
  and effective disinfection programs, particularly in smaller communities.
- Additional engineering research to simplify and lower the cost of removing
  excess nitrates and fluorides.
- Improved systems to control aesthetically undesirable concentrations of
  iron, manganese, hydrogen sulfide, and color, as well as taste and
  odor-causing organic constituents.
- Analytical surveillance techniques and control procedures to eliminate the
  deterioration that is occurring in water quality between the time the water
  leaves the treatment plant and the time it reaches the consumer.
- Improved planning to provide adequate quantities of safe water to the
  majority of our people who live in urban areas, and to assure optimum
  resource development and utilization to meet the needs of major population complexes.

History gives ample evidence of the inescapable penalties paid by past
civilizations which failed to provide for the safety of their drinking water
systems. Modern history shows that such waterborne diseases as typhoid,
dysentery, and cholera are controllable and, in fact, were all but eliminated in
the United States by the 1930s by applying the principles identified in the
Drinking Water Standards. This study demonstrates that we have begun to
backslide, which in turn, explains why it is that waterborne disease persists as
evidenced by the epidemic at Riverside, California in 1965 which affected
18,000 people, the 30 percent gastroenteritis attack rate in Angola, New York in
1968 due to a failure in the disinfection system, and the 60 percent infectious
hepatitis attack rate which afflicted the Holy Cross football team in 1969 as a
result of the ineffective cross-connection control procedures. These recent
episodes, reinforced by the findings of the current study, provide ample evidence
of the increasing potential for similar episodes unless we improve water system
operations consistent with currently accepted standards of practice.

We must also recognize numerous voids in existing technology which do not
allow measurement of the current effectiveness of existing procedures. The
current Drinking Water Standards do little more than mention viruses, neglect

numerous inorganic chemicals which are known to be toxic to man, and identify only one index that is supposed to cover the entire family of organic chemical compounds. These standards must be updated.

The need for knowledge about the health effects of waterborne contaminants is acute. Research is required, for example, to develop improved treatment control and surveillance procedures for viruses. The chronic long-term effects of chemical contaminants requires thorough investigation. For instance, we must determine the concentration levels at which numerous contaminants, such as mercury, molybdenium or selenium, cause adverse health effects. Similarly, we must mount a major attack on a host of synthetic organic chemicals which are growing at a rate of 500 new compounds per year. In addition to the threats posed by such well-publicized materials as pesticides, we now have to face a multitude of new organic chemical compounds. Recognizing our relatively fixed amount of ground and surface water supply, the increasing water needs of the general population and industry, and the need to reuse our available supplies to satisfy future demands, we can no longer afford to "wait and see what happens." We must begin to *investigate before we introduce* new compounds into the environment.

All this research is essential if we are to maintain at least the status quo for the current generation. These are issues confronting scientists and engineers today at all levels of government. But the overall water hygiene effort is this generation's responsibility to future generations. Indeed, answers to many of the currently identifiable research problems of today must be gained quickly if the current and future planners of our environment are to begin to formulate rational, economic and effective plans for the continued growth and development of our society.

Water Quality Criteria—
Public Water Supplies

Introduction

The National Technical Advisory Subcommittee on Public Water Supplies has found it necessary to make some rather arbitrary decisions in order to proceed with its task of developing raw water quality criteria for public water supplies. Because public water supplies commonly involve processing of the raw water to improve its quality before distributing it to consumers, and because treatment processes exist which can, at a price, convert almost any water including sea water and grossly polluted fresh water into a potable product, it is necessary to consider the type of treatment in any discussion of raw water quality criteria for public water supplies.

We have adopted as the considered treatment the most common processes in use in this country in their simplest form for the treatment of surface waters for public use. This may include coagulation (less than about 50 ppm alum, ferric sulfate, or copperas with alkali addition as necessary but without coagulant aids or activated carbon), sedimentation (6 hours or less), rapid sand filtration (3 gal/sq ft/min or less) and disinfection with chlorine (without consideration to concentration or form of chlorine residual). A wide variety of modifications of this basic treatment process are in use for removing various impurities or altering quality characteristics, but we have arbitrarily excluded these modifications in our deliberations because of the difficulty in deciding where to stop in considering the many modifications and elaborations of the basic process.

Definitions

We have listed two types of criteria defined as follows:

(a) *Permissible criteria.* Those characteristics and concentrations of substances in raw surface waters which will allow the production of a safe, clear, potable, aesthetically pleasing, and acceptable public water supply which meets the limits of Drinking Water Standards after

Reprinted from *Report of the Committee on Water Quality Criteria*, Federal Water Pollution Control Administration, 1968, pp. 18-26.

treatment. This treatment may include, but will not include more than, the processes described above.

(b) *Desirable criteria.* Those characteristics and concentrations of substances in the raw surface waters which represent high-quality water in all respects for use as public water supplies. Water meeting these criteria can be treated in the defined plants with greater factors of safety or at less cost than is possible with waters meeting permissible criteria.

Several words used in the table and in the text require explanation in order to convey the Subcommittee's intent:

Narrative. The presence of this word in the table indicates that the Subcommittee could not arrive at a single numerical value which would be applicable throughout the country for all conditions. Where this word appears, the reader is directed to the appropriate explanatory text.

Absent. The most sensitive analytical procedure in Standard Methods (9) (or other approved procedure) does not show the presence of the subject constituent.

Virtually absent. This terminology implies that the substance is present in very low concentrations and is used where the substance is not objectionable in these barely detectable concentrations.

Discussion

The subcommittee recognizes that surface waters are used for public water supply without treatment other than disinfection. Such waters at the point of withdrawal should meet Drinking Water Standards (10) in all respects other than bacterial quality.

It should be emphasized that many raw water sources which do not meet these permissible criteria have been and are being used to provide satisfactory public water supplies by suitable additions to and elaboration of the treatment processes defined above. In some instances, however, the water delivered to the customer is of marginal quality. Also the finished water is much more likely to become unsatisfactory if treatment plant irregularities occur. It is recognized that most of the surface water treatment plants providing water for domestic use in the United States are relatively small (7) and without sophisticated technical controls. Marginal quality characteristics, therefore, assume considerable importance to the managers of plants treating such supplies. This is the importance of the factors of safety mentioned in the definition of "desirable criteria." However, managers of all supplies would welcome improved raw water quality.

This Subcommittee believes that the criteria set forth herein can be used in setting standards of raw water quality only with a substantial amount of understanding and discretion. To a considerable extent this is related to the very great regional variations in water quality entirely aside from manmade pollution. In addition, human occupance and activity have inevitable effects on water quality. These facts make it difficult and sometimes impossible to develop uniform numerical criteria suitable for national application.

TABLE 1. Surface water criteria for public water supplies.

Constituent or characteristic	Permissible criteria	Desirable criteria	Paragraph
Physical:			
Color (color units)	75	<10	1
Odor	Narrative	Virtually absent	2
Temperature *	do	Narrative	3
Turbidity	do	Virtually absent	4
Microbiological:			
Coliform organisms	10,000/100 ml [1]	<100/100 ml [1]	5
Fecal coliforms	2,000/100 ml [1]	<20/100 ml [1]	5
Inorganic chemicals:	(mg/l)	(mg/l)	
Alkalinity	Narrative	Narrative	6
Ammonia	0.5 (as N)	<0.01	7
Arsenic *	0.05	Absent	8
Barium *	1.0	do	8
Boron *	1.0	do	9
Cadmium *	0.01	do	8
Chloride *	250	<25	8
Chromium,* hexavalent	0.05	Absent	8
Copper *	1.0	Virtually absent	8
Dissolved oxygen	≥4 (monthly mean) ≥3 (individual sample)	Near saturation	10
Fluoride *	Narrative	Narrative	11
Hardness *	do	do	12
Iron (filterable)	0.3	Virtually absent	8
Lead *	0.05	Absent	8
Manganese * (filterable)	0.05	do	8
Nitrates plus nitrites *	10 (as N)	Virtually absent	13
pH (range)	6.0–8.5	Narrative	14
Phosphorus *	Narrative	do	15
Selenium *	0.01	Absent	8
Silver *	0.05	do	8
Sulfate *	250	<50	8
Total dissolved solids * (filterable residue).	500	<200	16
Uranyl ion *	5	Absent	17
Zinc *	5	Virtually absent	8
Organic chemicals:			
Carbon chloroform extract * (CCE)	0.15	<0.04	18
Cyanide *	0.20	Absent	8
Methylene blue active substances *	0.5	Virtually absent	19
Oil and grease *	Virtually absent	Absent	20
Pesticides:			
Aldrin *	0.017	do	21
Chlordane *	0.003	do	21
DDT *	0.042	do	21
Dieldrin *	0.017	do	21
Endrin *	0.001	do	21
Heptachlor *	0.018	do	21
Heptachlor epoxide *	0.018	do	21
Lindane *	0.056	do	21
Methoxychlor *	0.035	do	21
Organic phosphates plus carbamates.*	0.1 [2]	do	21
Toxaphene *	0.005	do	8
Herbicides:			
2,4-D plus 2,4,5-T, plus 2,4,5-TP *	0.1	do	21
Phenols *	0.001	do	8
Radioactivity:	(pc/l)	(pc/l)	
Gross beta *	1,000	<100	8
Radium-226 *	3	<1	8
Strontium-90 *	10	<2	8

*The defined treatment process has little effect on this constituent.

[1] Microbiological limits are monthly arithmetic averages based upon an adequate number of samples. Total coliform limit may be relaxed if fecal coliform concentration does not exceed the specified limit.

[2] As parathion in cholinesterase inhibition. It may be necessary to resort to even lower concentrations for some compounds or mixtures. See par. 21

The criteria selected by the Subcommittee are listed in the table and discussed in the numbered paragraphs cited in the table. The paragraphs also include some rationale of the basis for the criteria. The fact that a substance is not included in these criteria does not imply that its presence is innocuous. It would be quite impracticable to prepare a compendium of all toxic, deleterious, or otherwise unwelcome agents that may enter a surface water supply.

Sampling

Sampling should be of such frequency and of such variety (time of day, season, temperature, river stage or flow, location, depth) as to properly describe the body of water designated for public water supply. Sampling should also be conducted in full cognizance of findings of the sanitary survey. Judgment should be exercised as to the relative desirability of frequent sampling at one point, such as the raw water intake, as compared to less frequent sampling at numerous locations, such as is required for stream profiles or cross sections.

It is clearly not possible to apply these criteria solely as maximum single sample values. The criteria should not be exceeded over substantial portions of time. If they are exceeded, efforts should be made to determine the cause, and corrective measures undertaken.

Analytical Methods

The criteria are based upon those analytical methods described in Standard Methods for the Examination of Water and Wastewater (9) or upon methods acceptable to water pollution control agencies.

Paragraph 1: Color A limit of 75 color units (platinum-cobalt standard) has been recommended to permit the defined plant to produce water meeting Drinking Water Standards (10) with moderate dosages of coagulant chemicals. At optimum pH the dosage usually required is linearly related to the color of the raw water, and higher color of the type commonly associated with swamp drainage and similar nonindustrial sources can be removed by increasing the coagulant dose. These criteria do not apply to colors resulting from dyes and some other industrial and processing sources which cannot be measured by the platinum-cobalt standard. Such colors should not be present in concentrations which cannot be removed by the defined method of treatment.

Paragraph 2: Odor. The effectiveness of the defined method of treatment in removing odorous materials from water is highly variable depending on the nature of the material causing the odor. For this reason, it has not been feasible to specify any permissible criterion in terms of threshold odor number. The raw water should not have objectionable odor. Any odors present should be removable by the defined treatment. It is desirable that odor be virtually absent.

Paragraph 3: Temperature. Surface water temperatures vary with geographical location and climatic conditions. Consequently no fixed criteria are feasible. However, any of the following conditions are considered to detract (sometimes seriously) from raw water quality for public water supply use:

(1) Water temperature higher than 85 F;

(2) More than 5 F water temperature increase in excess of that caused by ambient conditions;

(3) More than 1 F hourly temperature variation over that caused by ambient conditions;

(4) Any water temperature change which adversely affects the biota, taste, and odor, or the chemistry of the water;

(5) Any water temperature variation or change which adversely affects water treatment plant operation (for example, speed of chemical reactions, sedimentation basin hydraulics, filter wash water requirements, etc.);

(6) Any water temperature change that decreases the acceptance of the water for cooling and drinking purposes.

Paragraph 4: Turbidity. Turbidity in water must be readily removable by coagulation, sedimentation, and filtration; must not be present in quantities (either by weight or volume) that will overload the water treatment plant facilities; and must not cause unreasonable treatment costs. In addition, turbidity in water must not be frequently changing and varying in characteristics or in quantity to the extent that such changes cause upset in water treatment plant processing.

Customary methods for measuring and reporting turbidity do not adequately measure those characteristics harmful to public water supply and water treatment processing. A water with 30 Jackson turbidity units may coagulate more rapidly than one with 5 or 10 units. Similarly water with 30 Jackson turbidity units sometimes may be more difficult to coagulate than water with 100 units. Sometimes clay added to very low turbidity water will improve coagulation. Therefore, it has not been possible to establish a turbidity criterion in terms of Jackson turbidity units. Neither can a turbidity criterion be expressed in terms of mg/1 "undissolved solids" or "nonfilterable solids." The type of plankton, clay, or earth particles, their size and electrical charges, are far more determining factors than the Jackson turbidity units. Nevertheless, it must be clearly recognized that too much turbidity or frequently changing turbidity is damaging to public water supply.

The criterion for too much turbidity in water must relate to the capacity of the water treatment plant to remove turbidity adequately and continuously at reasonable cost. Water treatment plants are designed to remove the kind and quantity of turbidity to be expected in each water supply source. Therefore, any increase in turbidity and any fluctuating turbidity load over that normal to a water must be considered in excess of that permissible.

Paragraph 5: Coliform and Fecal Coliform Organisms. Bacteria have been used as indicators of sanitary quality of water since 1880 when *B. coli* and similar organisms were shown to be normal inhabitants of fecal discharges. The coliform group as presently recognized by Drinking Water Standards (*10*) is defined in Standard Methods for the Examination of Water and Wastewater (*9*). This group includes organisms that vary in biochemical and serologic characteristics and in their natural sources and habitats; i.e., feces, soil, water, vegetation, etc.

Because the sanitary significance of the various members of the coliform group derives from their natural sources, differentiation of fecal from nonfecal organisms is important to evaluate raw water quality (5). Fecal coliforms are characteristically inhabitants of warmblooded animal intestines. Members of other coliform subgroups may be found in soil, on plants and insects, in old sewage, and in waters polluted some time in the past.

The objective of using the coliform group as an indicator of the sanitary quality of water is to evaluate the disease-producing potential of the water. To estimate the probability of pathogens being contributed from feces, the coliform and fecal coliform content must be quantified.

In relation to raw water sources, the following suggestions are offered to help resolve some of the difficulties of data interpretation.

Fecal coliform organisms may be considered indicators of recent fecal pollution. It is necessary to consider all fecal coliform organisms as indicative of dangerous contamination. Moreover, no satisfactory method is currently available for differentiating between fecal organisms of human and animal origin.

In the absence of fecal coliform organisms, the presence of other coliform group organisms may be the result of less recent fecal pollution, soil runoff water, or, infrequently, fecal pollution containing only those organisms.

In general, the presence of fecal coliform organisms indicates recent and possibly dangerous pollution. The presence of other coliform organisms suggests less recent pollution or contributions from other sources of non-fecal origin.

In the past the coliform test has been the principal criterion of suitability of raw water sources for public water supply. The increase in chlorination of sewage treatment plant effluents distorts this criterion by reducing coliform concentrations without removing many other substances which the defined water treatment plant is not well equipped to remove. It is essential that raw water sources be judged as to suitability by other measures and criteria than coliform organism concentrations.

The defined water treatment plant is considered capable of producing water meeting Drinking Water Standards (10) bacteriological criteria from these limits. The difference between the suggested concentration of 10,000 coliforms per 100 ml and the erstwhile figure of 5,000 per 100 ml is justified by the difference between the Phelps Index and the MPN. The Subcommittee suggests these numbers and the additional consideration of fecal coliforms in order to provide more realistic parameters in full recognition of modern knowledge and not as a means of sanctioning increased bacterial pollution of waters destined for public water supply use.

Paragraph 6: Alkalinity. Alkalinity in water should be sufficient to enable floc formation during coagulation, must not be high enough to cause physiological distress in humans, and must be proper for a chemically balanced water (neither corrosive nor incrusting). A criterion for minimum and maximum alkalinity in public water supply is related to the relative amounts of bicarbonates, carbonates, and hydroxide ions causing the alkalinity; and also to the pH, filterable (dissolved) solids, and calcium content. Because the permissible criterion for filterable solids is 500 mg/1 and the pH range is 6.0 to 8.5, alkalinity should not be less than about 30 mg/1.

The criterion for maximum alkalinity cannot be expressed in calcium carbonate equivalents as determined from $0.02N$ H_2SO_4 titration because of the interrelationships stated above. However, alkalinity values higher than about 400 mg/l to 500 mg/l would be too high for public water supply use. Within the range of 30 mg/l to 500 mg/l, the alkalinity criterion should be that value which is normal to the natural water and which from experience is satisfactory for public water supply use. Frequent variations from normal values are detrimental to public water supply processing control.

Paragraph 7: Ammonia. Ammonia is a significant pollutant in raw water for public water supplies because its reactions with chlorine result in compounds with markedly less disinfecting efficiency than free chlorine. In addition, it is frequently an indicator of recent sewage pollution.

In the early days of waste treatment, the oxidation of ammonia to nitrates was one of the major objectives of waste treatment, but with the development of the BOD test, this objective became neglected. Greater attention to the design and operation of waste treatment plants for the oxidation of ammonia and organic nitrogen is needed to minimize the concentration of pollution forms in these receiving waters.

Paragraph 8: Arsenic, Barium, Cadmium, Chromium (Hexavalent), Copper, Chloride, Cyanide, Iron, Lead, Manganese, Phenols, Selenium, Silver, Sulfate, Zinc, and Radioactive Substances. The significance of these substances as contaminants of drinking water is discussed in Drinking Water Standards *(10)*. The permissible criteria in this report are those included in Drinking Water Standards. With the possible exception of iron and in some instances copper and zinc, the defined treatment plant does little or nothing to remove these substances.

Paragraph 9: Boron. Boron is found in the natural ground and surface waters in some areas of the United States, notably in the Western States where as much as 5 to 15 mg/l are encountered. However, extensive data on boron in both well and surface waters in North America show that the amount of boron normally encountered is less than 1 mg/l. The ingestion of large amounts of boron can affect the central nervous system and protracted ingestion may result in a clinical syndrome known as borism.

Boron is an essential element to plant growth but is toxic to many plants at levels as low as 1 mg/l. The Public Health Service has established a limit of 1 mg/l which provides a good factor of safety physiologically and also considers the domestic use of water for home gardening.

Paragraph 10: Dissolved Oxygen. Criteria for dissolved oxygen are included, not because the substance is of appreciable significance in water treatment or in finished water, but because of its use as an indicator of pollution by organic wastes. It is intended for application to freeflowing streams and not to lakes or reservoirs in which supplies may be taken from below the thermocline.

Paragraph 11: Fluoride. The Subcommittee recognizes the potential ·beneficial effects of fluoride ion in domestic water supplies but recommends no "desirable" concentration since any value less than that recommended for the permissible limit would be acceptable from the point of view of a water pollution control program. Rapid fluctuations in raw-water fluoride ion levels would create objectionable operating problems for communities supplementing raw-water fluoride concentrations. The permissible criterion is the upper limit of the recommended range in Drinking Water Standards (*10*).

Annual average of maximum daily air temperatures[1] F:	Recommended Limit mg/l
50.0 to 53.7	1.7
53.8 to 58.3	1.5
58.4 to 63.8	1.3
63.9 to 70.6	1.2
70.7 to 79.2	1.0
79.3 to 90.5	0.8

[1] Based on temperature data obtained for a minimum of 5 years.

Paragraph 12: Hardness. A singular criterion for the maximum hardness in public water supply is not possible. Hardness in water is largely the result of geological formations with which the water comes in contact. Public acceptance of hardness varies from community to community. Consumer sensitivity to objectionable hardness is related to the hardness with which he has become accustomed. Consumer acceptance of hardness may also be tempered by economic necessity.

Hardness should not be present in concentrations that will cause excessive soap consumption, or which will cause objectionable scale in heating vessels and pipes. In addition, varying water hardness is objectionable to both domestic and industrial water consumers. With varying hardness, the soap required for laundry, the effect on manufactured products, and the damage to process equipment (such as boilers and cooling coils) cannot be anticipated and compensated without facilities which are not available to most water users. A water hardness criterion must relate to the hardness which is normal to the supply and exclude hardness additions which will cause variations.

A criterion for objectionable hardness must be tailored to fit the requirements of each community. Hardness more than 300-500 mg/l as $CaCO_3$ is excessive for public water supply. Many consumers will object to water harder than 150 mg/l. In other communities, the criterion for maximum water hardness is considerably less than 150mg/l. A moderately hard water is sometimes defined as having hardness between 60 to 120 mg/l.

Paragraph 13: Nitrate plus Nitrite. A limit of 10mg/l (N) of nitrate ion plus nitrite ion will be recommended by Drinking Water Standards (*10*). Because the nitrite ion is the substance actually responsible for causing methemoglobinemia, a combined limit on the two ions is more significant than a limit on nitrates only.

Paragraph 14: pH. Most unpolluted waters have pH values within the range recommended as a permissible criterion. Any pH value within this range is acceptable for public water supply. The further selection of a specific pH value within this range as a desirable criterion cannot be made.

Paragraph 15: Phosphorus. The Subcommittee has considered establishing criteria on phosphorus concentrations but has not been able to establish any generally acceptable limit because of the complexity of the problem. The purpose of such a limit would be twofold:

(a) To avoid problems associated with algae and other aquatic plants, and
(b) To avoid coagulation problems due particularly to complex phosphates.

Phosphorus is an essential element for aquatic life as well as for all forms of life and has been considered the most readily controllable nutrient in efforts to limit the development of objectionable plant growths. Evidence indicates that high phosphorus concentrations are associated with the eutrophication of waters that is manifest in unpleasant algal or other aquatic plant growths when other growth-promoting factors are favorable; that aquatic plant problems develop in reservoirs or other standing waters at phosphorus values lower than those critical in flowing streams; that reservoirs and other standing waters will collect phosphates from influent streams and store a portion of these within the consolidated sediments; that phosphorus concentrations critical to noxious plant growths will vary with other water quality characteristics, producing such growths in one geographical area but not in another.

Because the ratio of total phosphorus to that form of phosphorus readily available for plant growth is constantly changing and will range from two to 17 times or greater, it is desirable to establish limits on the total phosphorus rather than the portion that may be available for immediate plant use. Most relatively uncontaminated lake districts are known to have surface waters that contain 10 to 30 μg/l total phosphorus as P; in some waters that are not obviously polluted, higher values may occur (4). Data collected by the Federal Water Pollution Control Administration, Division of Pollution Surveillance, indicate that total phosphorus concentrations exceeded 50 μg/l (P) at 48 percent of the stations sampled across the Nation (6). Some potable surface water supplies now exceed 200 μg/l (P) without experiencing notable problems due to aquatic growths. Fifty micrograms per liter of total phosphorus (as P) would probably restrict noxious aquatic plant growths in flowing waters and in some standing waters. Some lakes, however, would experience algal nuisances at and below this level.

Critical phosphorus concentrations will vary with other water quality characteristics. Turbidity and other factors in many of the Nation's waters negate the algal-producing effects of high phosphorus concentrations. When waters are detained in a lake or reservoir, the resultant phosphorus concentration is reduced to some extent over that in influent streams by precipitation or uptake by organisms and subsequent deposition in fecal pellets or dead organism bodies. See the report of the Subcommittee for Fish, Other Aquatic Life, and Wildlife, and the section on Plant Nutrients and Nuisance Organisms for a more complete discussion of phosphorus associations with the enrichment problem.

At concentrations of complex phosphates of the order of 100μg/l, difficulties with coagulation are experienced.

Paragraph 16: Total Dissolved Solids (Filterable Residue). Drinking Water Standards (10) recommend that total dissolved solids not exceed 500 mg/l where other more suitable supplies are available. It is noted, however, that some streams contain total dissolved solids in excess of 500 mg/l. For example, the Colorado River at the point of withdrawal by the Metropolitan Water District of Southern California has a total dissolved solids concentration up to 700 mg/l.

High total dissolved solids are objectionable because of physiological effects, mineral taste, or economic effect. High concentrations of mineral salts, particularly sulfates and chlorides, are associated with corrosion damage in water systems. Regarding taste, on the basis of limited research work underway in California, limits somewhat higher than 500 mg/l are probably acceptable to consumers of domestic water supplies. It is likely that levels set with relation to economic effects are controlling for this parameter.

Increases in total dissolved solids from those normal to the natural stream are undesirable and may be detrimental.

It is recommended that the permissible value for total dissolved solids be set at 500 mg/l in view of the above evaluation. Further, it is recommended that research work be sponsored to obtain more information on total dissolved solids in water relating to physiological effects, consumer attitudes toward taste, and economic considerations.

Paragraph 17: Uranyl Ion. The standard for uranyl ion (UO_2=) is established on the basis of its chemical properties rather than on the basis of its being a radioactive material. It is being added to Drinking Water Standards (*10*). Uranyl ion is of concern in drinking water because of possible damage to the kidneys. The threshold level of taste and the appearance of color due to uranyl ion occur at about 10 mg/l which is much less than the safe limit of ingestion of this ion insofar as adverse physiological effects are concerned.

The Public Health Service adopted the figure of 5 mg/l which is one-half the limit based on taste and color and, therefore, there is a considerable factor of safety in the adoption of 5 mg/l.

Paragraph 18: Carbon Chloroform Extract (CCE). A limit of 0.2 mg/l carbon chloroform extract in drinking water is recommended in the Drinking Water Standards "as a safeguard against the intrusion of excessive amounts of potentially toxic material into water" (*10, 3*). Although the analytical procedure then in use leaves much to be desired from the standpoint of simplicity, reproducibility, and interpretation, it was the best available at that time. The analytical procedure has been improved since then and the newer technique (*1, 2*) gives substantially higher results than the one originally used. The defined method of treatment generally removes very little of the CCE present in the raw water. In many instances there is an increase during treatment. Whether this is real or apparent is not known.

The permissible criterion of 0.15 mg/l recommended is based on use of the procedure cited in Drinking Water Standards (*10*). We do not as yet have sufficient information on which to base a recommended limit using the lower flow rates and sample volumes of the newer procedure. When this information is available, a change in the criterion is advisable. This limit is generally attainable where vigorous efforts at pollution control are carried out.

Paragraph 19: Methylene Blue Active Substances. This is an operationally more precise name for substances discussed in Drinking Water Standards (10) as alkyl benzene sulfonate. The permissible criterion is the same as the limit recommended in those standards. Those standards have been revised to reflect this change in nomenclature.

Paragraph 20: Oil and Grease. It is very important that water for public water supply be free of oil and grease. The difficulty of obtaining representative samples of these materials from water makes it virtually impossible to express criteria in numerical units. Since even very small quantities of oil and grease may cause troublesome taste and odor problems, the Subcommittee desires that none of this material be present in public water supplies. An additional problem attributable to these agents is the unsightly scumlines on water treatment basin walls, swimming pools, and other containers.

Paragraph 21: Pesticides and Herbicides. Consideration was given by the Subcommittee to three groups of pesticides: the more common chlorinated hydrocarbons, herbicides, and the cholinesterase-inhibiting group which include the organic phosphorus types and the carbamates. The permissible levels are based upon recommendations of the Public Health Service Advisory Committee on Use of the PHS Drinking Water Standards. These values were derived for that Committee by an expert group of toxicologists as those levels which, if ingested over extensive periods, could not cause harmful or adverse physiological changes in man. In the case of aldrin, heptachlor, chlordane, and parathion, the Committee adopted even lower than physiologically safe levels; namely, amounts which, if present, can be detected by their taste and odor. It should be noted that this National Technical Advisory Subcommittee on Public Water Supplies is not a group of toxicological experts. Hence, the promulgation of additional criteria by the Public Health Service would also be accommodated by this Subcommittee, tempered—as was done above—by its experience and judgement in the area of water treatment, as, for example, in public acceptance of organoleptic properties.

The limit for the cholinergic pesticides is established relative to parathion and is expressed as 0.1 mg/l parathion equivalent. This equivalence is the ratio that a given pesticide of this group has to parathion as unity in its cholinesterase inhibiting properties. This makes it incumbent upon an administrator of this limit to determine the pesticide involved and to obtain expert toxicological opinion on its parathion equivalence. Nearly all the organophosphorus compounds and the cholinergic carbamates have high acute toxicity to mammals and some have even higher toxicity to fish. Ingestion of small quantities of these compounds over long time periods causes damage to mammalian central nervous systems. Many organophosphorus pesticides hydrolyze rapidly in the environment to harmless or less harmful products. The hazards from the chlorinated hydrocarbon pesticides in water results from both direct effects, because they tend to persist in their original form over long periods, and indirect effects because they may be concentrated biologically in man's food chain. The values which were selected by the Public Health Service as limits for this group of pesticides are, however, set with substantial safety factors insofar as they adversely affect the human body. Generally, fish are more sensitive to this group of pesticides and, therefore, may serve as a rough method for determining when the chlorinated hydrocarbon pesticides content of water is approaching a danger level. See the report of the Fish, Other Aquatic Life, and Wildlife Subcommittee for pesticide limits relative to maintaining healthy and productive aquatic life.

It should be noted that limits for pesticides and herbicides have been set with relation only to human intake directly from a related domestic water supply. The consequence of higher and possibly objectionable concentrations in fish available to be eaten by man due to biological concentration is considered not within the scope of the charge to this Subcommittee.

References

1. Booth, R. L., J. N. English, and G. N. McDermott. 1965. Evaluation of sampling conditions in the carbon adsorption method. J. Amer. Water Works Assoc. 57: 215-220.

2. Breidenbach, A. W., et al. 1966. The identification and measurement of chlorinated hydrocarbon pesticides in surface waters. WP–22, U.S. Department of the Interior, Federal Water Pollution Control Administration, Washington, D.C.

3. Ettinger, M. B. 1960. A proposed toxicological screening procedure for use in water works. J. Amer. Water Works Assoc. 52: 689-694.

4. Gales, M. F., Jr., E. C. Julian, and R. C. Kroner. 1966. Method for quantitative determination of total phosphorus in water. J. Amer. Water Works Assoc. 58: 1363–1368.

5. Geldreich, E. E. 1966. Sanitary significance of fecal coliforms in the environment. U.S. Department of Interior, Federal Water Pollution Control Administration, Washington, D.C.

6. Gunnerson, C. B. 1966. An atlas of water pollution surveillance in the United States, Oct. 1, 1957, to Sept. 30, 1965. U.S. Department of the Interior, Federal Water Pollution Control Administration, Cincinnati, Ohio.

7. Koenig, L. 1967. The cost of water treatment by coagulation, sedimentation, and rapid sand filtration. J. Amer. Water Works Assoc. 59: 290-336.

8. Middleton, F. M., A. A. Rosen, and R. H. Burttschell. 1962. Tentative method for carbon chloroform extract (CCE) in water. J. Amer. Water Works Assoc. 54: 223-227.

9. Standard Methods for the Examination of Water and Wastewater. 1967. 12th ed. New York: Amer. Public Health Assoc.

10. U.S. Department of Health, Education, and Welfare, Public Health Service 1962. Drinking water standards. Public Health Service Pub. 956.

Constituents and Properties of Water

C. J. Durfor* and Edith Becker**

After a long, hard hike through the woods, a sip of cool spring water tastes wonderful; at such a time, it is difficult to think about the chemicals in water. One is tempted to say that this spring water is ideal—after all, it is clear, cool, and sparkling, and it tastes good. But is it ideal water?

This same water may contain minute amounts of chemicals that can cause bodily harm if the water is ingested over a long period of time. It may contain chemical constituents that will consume large amounts of soap and detergents, or it may contain constituents that will stain porcelain fixtures and laundry. It may contain constituents that will form scale, which will gradually choke pipes, or it may lack these same constituents, and then it will corrode pipes. Does this mean we should keep away from water? No! What we are saying is that all natural water contains chemical constituents—from A (aluminum) to Z (zinc)—and that the amounts of these constituents vary from too much for one purpose to too little for another purpose; as a result, water generally has to be treated before it can be used.

Major Chemical Constituents

The chemical constituents most commonly found in water are silica, iron, manganese, calcium, magnesium, sodium, potassium, carbonate, bicarbonate, sulfate, chloride, fluoride, nitrate, and dissolved solids (dissolved solids is the residue after evaporating a water sample at 180°C). . . .

Table 1 shows the major sources of each of these constituents and also the maximum concentrations of these constituents in surface and ground water, ocean water, and natural brines. In unpolluted surface and ground water, the

Reprinted from Public Water Supplies of the 100 Largest Cities in the United States, 1962; U.S. Geol. Survey Water Supply Paper 1812, 1968, pp. 14-32. This report has been slightly edited.
*Water Resources Division, U.S. Geological Survey, Washington, D.C.
**Water Resources Division, U.S. Geological Survey, Washington, D.C.

24

occurrence and amount of these constituents are regulated to a large extent by the geologic environment. In Pennsylvania, for example, the headwaters of a certain stream originating in an anthracite coal field are laden with sulfate; the pH is less than 4.0. When the stream leaves the anthracite field it contains no bicarbonate; it then flows through an area underlain with limestone, and here the sulfate content decreases, the bicarbonate content increases, and pH increases to more than 4.5. Near its mouth this stream passes through a limestone quarry; here the stream picks up more bicarbonate, and the pH becomes greater than 9.0.

Most natural water contains calcium and magnesium; these elements are known as the alkaline earths and are the chief cations found in many waters. (An ion is an element, or a group of elements combined to act as a single constituent, that has an electrical charge; an ion with a negative charge is an anion; an ion with a positive charge is a cation.) It is not uncommon for natural water to contain several times as much calcium as magnesium.

Sodium and potassium are common alkali metals found in water; generally, they are present in much smaller quantities than the alkaline earths. In southern Louisiana and Texas, calcium and magnesium in ground water are exchanged with sodium and potassium in the soil, and the resultant water is enriched in sodium and potassium and contains negligible amounts of calcium and magnesium. Streams receiving waste water from irrigation and streams in arid areas, in tidal areas, and in areas underlain by sodium chloride beds also contain considerably more alkali metals than alkaline earths.

Carbonate and bicarbonate are found in most natural water because of the abundant deposits of readily soluble limestone (composed principally of calcium carbonate) and dolomite (composed principally of magnesium and calcium carbonates). In the presence of carbon dioxide, the dissolving of carbonate rocks by water forms anions of bicarbonates and carbonates in water.... Many ground-water environments are favorable to the dissolving of limestone rocks and so large amounts of bicarbonate are present in the water. In different environments, water saturated with calcium carbonate may reprecipitate calcium carbonate.

Sulfate is present in natural water but is commonly not found in as large an amount as is bicarbonate; however, water draining mining areas, gypsum beds, and arid lands frequently contains more sulfate than bicarbonate.

Chloride and nitrate are commonly found in all water, generally in amounts less than 10 ppm.

For more than 60 years the dental defect that appears as a dark stain on tooth enamel and that is known as mottled enamel—locally called "Texas teeth" or "Colorado stain"—has been under investigation. It was reasoned that this defect was caused by some trace element in water. Later, fluorine was proved to be the cause (McNeil, 1957). Still later, fluoride concentrations of about 0.6 to 1.7 ppm in water were found to reduce the incidence of dental caries, and concentrations greater than 1.7 ppm were found to protect the teeth from cavities but to cause an undesirable black stain (U.S. Public Health Service, 1962b). For further information on the physiological effects of fluoride, the

TABLE 1. Major chemical constituents in water—their sources, concentrations, and effects upon usability.

[Concentrations are in parts per million. Table prepared with the assistance of B. P. Robinson]

Constituent	Major sources	Concentration in natural water	Effect upon usability of water	Concentration in public water supplies of the 100 cities				Drinking water should contain less than concentration shown if more suitable supplies are or can be made available [1]
				Untreated water		Treated water		
				Max.	Min.	Max.	Min.	
Silica (SiO$_2$)	Feldspars, ferromagnesium and clay minerals, amorphous silica cachert, opal.	Ranges generally from 1.0 to 30 ppm, although as much as 100 ppm is fairly common; as much as 4,000 ppm is found in brines.	In the presence of calcium and magnesium, silica forms a scale in boilers and on steam turbines that retards heat; the scale is difficult to remove. Silica may be added to soft water to inhibit corrosion of iron pipes.	72	0.0	72	0.0	
Iron (Fe)	1. *Natural sources:* Igneous rocks: Amphiboles, ferromagnesian micas, ferrous sulfide (FeS), ferric sulfide (FeS), ferrous sulfide or iron pyrite (FeS$_2$), magnetite (Fe$_3$O$_4$). Sandstone rocks: Oxides, carbonates, and sulfides of iron clay minerals. 2. *Manmade sources:* Well casing, piping, pump parts, storage tanks, and other objects of cast iron and steel which may be in contact with the water. Industrial wastes.	Generally less than 0.50 ppm in fully aerated water. Ground water having a pH less than 8.0 may contain 10 ppm; rarely as much as 50 ppm may occur. Acid water from thermal springs, mine wastes, and industrial wastes may contain more than 6,000 ppm.	More than 0.1 ppm precipitates after exposure to air; causes turbidity, stains plumbing fixtures, laundry and cooking utensils, and imparts objectionable tastes and colors to foods and drinks. More than 0.2 ppm is objectionable for most industrial uses.	1.90	0.00	1.30	0.00	0.3
Manganese (Mn)	Manganese in natural water probably comes most often from soils and sediments. Metamorphic and sedimentary rocks and mica biotite and amphibole hornblende minerals contain large amounts of manganese.	Generally 0.20 ppm or less. Ground water and acid mine water may contain more than 10 ppm. Reservoir water that has "turned over" may contain more than 150 ppm.	More than 0.2 ppm precipitates upon oxidation; causes undesirable tastes, deposits on foods during cooking, stains plumbing fixtures and laundry, and fosters growths in reservoirs, filters, and distribution systems. Most industrial users object to water containing more than 0.2 ppm.	0.60	0.00	2.5	0.00	0.05

Constituent	Source	Concentration in water	Effects or significance				
Calcium (Ca)	Amphiboles, feldspars, gypsum, pyroxenes, aragonite, calcite, dolomite, clay minerals.	As much as 600 ppm in some western streams; brines may contain as much as 75,000 ppm.	Calcium and magnesium combine with bicarbonate, carbonate, sulfate, and silica to form heat-retarding, pipe-clogging scale in boilers and in other heat-exchange equipment. Calcium and magnesium combine with ions of fatty acid in soaps to form soap suds; the more calcium and magnesium, the more soap required to form suds. A high concentration of magnesium has a laxative effect, especially on new users of the supply.	145	0.0	145	0.0
Magnesium (Mg)	Amphiboles, olivine, pyroxenes, dolomite, magnesite, clay minerals.	As much as several hundred parts per million in some western streams; ocean water contains more than 1,000 ppm, and brines may contain as much as 57,000 ppm.		120	0.0	120	0.0
Sodium (Na)	Feldspars (albite); clay minerals; evaporites, such as halite (NaCl) and mirabilite (Na$_2$SO$_4$·10H$_2$O); industrial wastes.	As much as 1,000 ppm in some western streams; about 10,000 ppm in sea water; about 25,000 ppm in brines.	More than 50 ppm sodium and potassium in the presence of suspended matter causes foaming, which accelerates scale formation and corrosion in boilers. Sodium and potassium carbonate in recirculating cooling water can cause deterioration of wood in cooling towers. More than 65 ppm of sodium can cause problems in ice manufacture.	177	1.1	198	1.1
Potassium (K)	Feldspars (orthoclase and microcline), feldspathoids, some micas, clay minerals.	Generally less than about 10 ppm; as much as 100 ppm in hot springs; as much as 25,000 ppm in brines.		30	0.2	30	0.0
Carbonate (CO$_3$)	Limestone, dolomite.	Commonly 0 ppm in surface water; commonly less than 10 ppm in ground water. Water high in sodium may contain as much as 50 ppm of carbonate.	Upon heating, bicarbonate is changed into steam, carbon dioxide, and carbonate. The carbonate combines with alkaline earths—principally calcium and magnesium—to form a crustlike scale of calcium carbonate that retards flow of heat through pipe walls and restricts flow of fluids in pipes. Water containing large amounts of bicarbonate and alkalinity are undesirable in many industries.	21	0	26	0
Bicarbonate (HCO$_3$)	Limestone, dolomite.	Commonly less than 500 ppm; may exceed 1,000 ppm in water highly charged with carbon dioxide.		380	5	380	0
Sulfate (SO$_4$)	Oxidation of sulfide ores; gypsum; anhydrite; industrial wastes.	Commonly less than 1,000 ppm except in streams and wells influenced by acid mine drainage. As much as 200,000 ppm in some brines.	Sulfate combines with calcium to form an adherent, heat-retarding scale. More than 250 ppm is objectionable in water in some industries. Water containing about 500 ppm of sulfate tastes bitter; water containing about 1,000 ppm may be cathartic.	572	0.0	572	0.0	250

[1] U.S. Public Health Service (1962b).

TABLE 1. Major chemical constituents in water—their sources, concentrations, and effects upon usability—Continued.

[Concentrations are in parts per million. Table prepared with the assistance of B. P. Robinson]

Constituent	Major sources	Concentration in natural water	Effect upon usability of water	Concentration in public water supplies of the 100 cities				Drinking water should contain less than concentration shown if more suitable supplies are or can be made available [1]
				Untreated water		Treated water		
				Max.	Min.	Max.	Min.	
Chloride (Cl)	Chief source is sedimentary rock (evaporites); minor sources are igneous rocks. Ocean tides force salty water upstream in tidal estuaries.	Commonly less than 10 ppm in humid regions; tidal streams contain increasing amounts of chloride (as much as 19,000 ppm) as the bay or ocean is approached. About 19,300 ppm in sea water; and as much as 200,000 ppm in brines.	Chloride in excess of 100 ppm imparts a salty taste. Concentrations greatly in excess of 100 ppm may cause physiological damage. Food processing industries usually require less than 250 ppm. Some industries—textile processing, paper manufacturing, and synthetic rubber manufacturing—desire less than 100 ppm.	540	0.5	540	0.0	250
Fluoride (F)	Amphiboles (hornblende), apatite, fluorite, mica.	Concentrations generally do not exceed 10 ppm in ground water or 1.0 ppm in surface water. Concentrations may be as much as 1,600 ppm in brines.	Fluoride concentration between 0.6 and 1.7 ppm in drinking water has a beneficial effect on the structure and resistance to decay of children's teeth. Fluoride in excess of 1.5 ppm in some areas causes "mottled enamel" in children's teeth. Fluoride in excess of 6.0 ppm causes pronounced mottling and disfiguration of teeth.	7.0	0.0	7.0	0.0	The recommended control limits depend upon annual averages of maximum daily air temperature and range from 0.6 to 0.8 ppm at 79.3° to 90.5° F and 0.9 to 1.7 ppm at 50.0° to 53.7° F.

| Nitrate (NO_3) | Atmosphere; legumes, plant debris, animal excrement, nitrogenous fertilizer in soil and sewage. | In surface water not subjected to pollution, concentration of nitrate may be as much as 5.0 ppm but is commonly less than 1.0 ppm. In ground water the concentration of nitrate may be as much as 1,000 ppm. | Water containing large amounts of nitrate (more than 100 ppm) is bitter tasting and may cause physiological distress. Water from shallow wells containing more than 45 ppm has been reported to cause methemoglobinemia in infants. Small amounts of nitrate help reduce cracking of high-pressure boiler steel. | 23 | 0.0 | 23 | 0.0 | 45 (In areas in which the nitrate content of water is known to be in excess of the listed concentration, the public should be warned of the potential dangers of using the water for infant feeding.) |
| Dissolved solids | The mineral constituents dissolved in water constitute the dissolved solids. | Surface water commonly contains less than 3,000 ppm; streams draining salt beds in arid regions may contain in excess of 15,000 ppm. (Ground water commonly contains less than 5,000 ppm; some brines contain as much as 300,000 ppm. | More than 500 ppm is undesirable for drinking and many industrial uses. Less than 300 ppm is desirable for dyeing of textiles and the manufacture of plastics, pulp paper, rayon. Dissolved solids cause foaming in steam boilers; the maximum permissible content decreases with increases in operating pressure. | 1,580 | 10 | 1,580 | 22 | 500 |

[1] U.S. Public Health Service (1962b).

reader is referred to a selection of papers on the subject prepared by the U.S. Public Health Service (1962a).

Water boiled in a dish leaves a crust of salt composed principally of silica, calcium, magnesium, sodium, potassium, bicarbonate, carbonate, sulfate, chloride, nitrate, and some water bound in the residue. Upon heating this residue to $180°C$, two changes occur: most of the water of crystallization is expelled, and most bicarbonate is converted to carbonate. The residue dried at $180°C$ (called residue on evaporation) approximates the quantity of anhydrous chemicals in solution and is used as an indication of the dissolved-solids content in the water.

In many locations, efforts are made to obtain an adequate public supply of water containing small amounts of dissolved solids; the cost of treating water generally increases with increased amounts of dissolved solids. However, a person accustomed to drinking water containing a moderate amount of dissolved solids may complain about the "flat taste" of drinking water that has less than 100 ppm of dissolved solids. The amount of dissolved solids in the untreated water used for public supply ranges from less than 100 ppm along the Appalachian Mountains and in the far West to more than 500 ppm in the arid Southwest. (See fig. 1).

Many of the largest cities obtain their water supplies from more than one source. For these cities, the dissolved-solid contents were weighted in proportion to the population served by each water source (fig. 1). The dissolved-solids content of each source was multiplied by the population served by that source. The products of the dissolved solids and population for each source were added. The resultant sum of the products was divided by the population served by all sources in the city to obtain a population-weighted average dissolved-solids content.

Many of the calculations of the population-weighted dissolved-solids content are based upon yearly averages supplied by officials of city waterworks. Other calculations of population-weighted dissolved-solids content are based on samples collected so as to represent an average value. The dissolved-solids content for a few cities was not calculated because of a lack of data.

The presence of specific amounts of certain constituents can have an adverse effect upon the usability of water. A few of the known tolerances of specific chemicals that affect the usability of water are listed in table 1. Some constituents, such as iron and manganese, are detrimental, even in small quantities. Fortunately, most water used by industry—more than 95 percent—is used for cooling, for which the main prerequisites are that the water be free of sediment, debris, and algae that could clog pipes. For a more comprehensive report on "quality tolerance of water for industrial uses," the reader is referred to Moore (1940).

Since about 1914, criteria have been promulgated to govern the quality of drinking water used on interstate carriers. The drinking-water standards established by the U.S. Public Health Service have gained wide acceptance and are now used by many water authorities as a guide in determining local drinking-water standards. These standards provide two types of chemical limits:

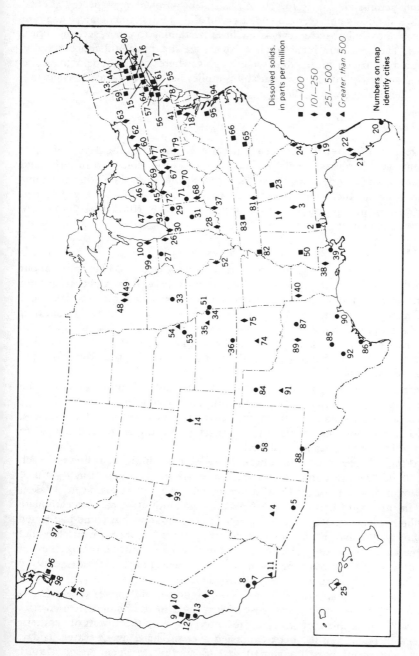

FIGURE 1. Dissolved solids in untreated public water supplies of the 100 largest cities in the United States, 1962. (Average weighted by population served.)

maximum permissible limits for chemicals having known or suspected adverse physiological effects and recommended permissible limits for chemicals that are generally nontoxic but have adverse qualities pertaining to color, staining, taste, and odor. The concentrations shown in table 1 are the maximum concentrations that should be found in a public water supply where "in the judgment of the certifying authority, other more suitable supplies are or can be made available" (U.S. Public Health Service, 1962b).

Major Properties

The properties of water that influence the use of water and the degree of water treatment are hardness, specific conductance, pH, color, turbidity, and temperature. The description and causes of these properties, their concentrations in natural water, the effect of concentrations upon usability of water, and concentrations in the public water supplies of the 100 largest cities are summarized in table 2.

Hardness

In one part of the country, a newcomer may be provoked into exclaiming about the hardness of the water because of his difficulty in working up a lather with soap and water. In another part of the country, a newcomer may remark about the softness of the water because soap suds are so easily formed. Hardness of water is a property of water that is a measure of the amount of soap required to form a lather. Not too many years ago the hardness of water was measured in the laboratory by determining the amount of soap solution that must be added to water to form suds.

Before soap can form a lather, part of the soap molecule must react with the calcium and magnesium in the water to form an insoluble curd. The smaller the amounts of calcium and magnesium, the easier soap suds are formed; conversely, the greater the amounts of calcium and magnesium, the more soap curds are formed and the more soap is consumed.

In 1856, Thomas Clark, of England, defined hardness as follows: "Each degree of hardness is as much as a grain of chalk or the calcium in a grain of chalk would produce in a gallon of water, by whatever means dissolved" (Baker, 1948). In this report hardness is expressed as the amount of calcium carbonate in a million parts of water chemically equal to the amount that could be formed from the calcium and magnesium in solution. Aluminum, iron, manganese, and other metals in water also consume soap and thus contribute to the hardness of water; however, the amounts in which they are present in the water are generally small, and their effect upon hardness is insignificant.

In 1933 when soap—not detergent—was a household name, it was firmly established that the amount and cost of soap used in the home increased with increases in the hardness of water. In recent years, with the advent of synthetic detergents, less concern has been expressed over the hardness of water. Today, synthetic detergents outsell soaps 10 to 1 (Soap and Detergent Assoc., 1962), and some people think that synthetic detergents are as effective in hard water as in soft water. However, most synthetic detergents contain about 30—50 percent sequestering ingredients that react with calcium and magnesium, the hardness components of water. "In hard water these ingredients are decreased in effective concentration for their cleaning purpose" (DeBoer and Larson, 1961). A recent study indicated that three times the amount of synthetic detergents was required for 400 ppm hardness water than for 0 ppm hardness (Aultman, 1957).

Sixteen cities, serving more than 15 million people, have "moderately hard" (61—120 ppm) raw water and do not soften their supply; laundries and other industries consider it advantageous to remove some of the hardness. Many municipalities try to reduce the hardness of their water supply to 85—100 ppm.

Twenty-two cities, serving almost 16 million people, have "hard" (121—180) ppm) raw water for their public supply. Homes using hard water have more problems with soap curds than homes that use softer water. Many industries require that "hard" water be treated to lower the hardness. About one-half of the 22 cities, serving about 6 million people, lower the hardness by some type of water softening.

Twenty-seven cities, supplying more than 8 million people, have "very hard" (more than 180 ppm) raw water; only 15 of these cities, serving more than 5 million people, lower the hardness.

The anions in water—principally bicarbonate and carbonate—determine the proportions of "carbonate" and "noncarbonate" hardness that constitute the hardness of water. Carbonate hardness is the amount of hardness chemically equivalent to the amount of bicarbonate and carbonate in solution. Carbonate hardness is approximately equal to the amount of hardness that is removed from water by boiling. Carbonate hardness of water results in the deposition of a calcium and magnesium carbonate scale, especially at temperatures above boiling point; this scale impedes the transfer of heat and constricts the effective pipe diameter, which reduces the flow of water.

Noncarbonate hardness is the difference between the hardness calculated from the total amount of calcium and magnesium in solution and the carbonate hardness. If the carbonate hardness (expressed as calcium carbonate) equals the amount of calcium and magnesium hardness (also expressed as calcium carbonate), there is no noncarbonate hardness. Noncarbonate hardness is about equal to the amount of hardness remaining after water is boiled. The scale formed at high temperatures by the evaporation of water containing non-carbonate hardness is tough, heat resistant, and difficult to remove.

Soft water and hard water are common terms, but there is no clear line of demarcation. Water that seems hard to an easterner may seem soft to many westerners. The hardness-of-water classification used in his report is as follows:

Hardness range (parts per million of calcium carbonate)	*Hardness description*
0—60	Soft.
61—120	Moderately hard.
121—180	Hard.
More than 180	Very Hard.

TABLE 2. Properties of water—their description, concentrations, and effects upon usability of water.

Property and unit of measurement	Description and causes	Concentration or value in natural water	Effect upon usability of water	Concentration or value in public water supplies of the 100 largest cities			
				Untreated water		Treated water	
				Max.	Min.	Max.	Min.
Hardness (parts per million)	Hardness is expressed as the quantity of calcium carbonate equivalent to the calcium and magnesium present. It is caused principally by calcium and magnesium ions, but other alkaline earths (barium and strontium) and free acid and heavy-metal ions contribute to hardness.	In most surface water the hardness is less than 1,000 ppm; in ground water, the hardness is generally less than 2,000 ppm. In arid regions the hardness of surface and ground waters may be higher.	Water low in hardness causes corrosion of metallic surfaces. Hard water consumes excessive amounts of soap and synthetic detergents in homes, laundries, and textile industries; it forms insoluble scum and curds and causes problems in the processing of foods, beverages, and rubber. Hardness of water is classified as follows: Hardness range *Description* (ppm) 0–60 .... Soft. 61–120 .... Moderately hard. 121–180 .... Hard. More than 180 .... Very hard.	738	0	738	0
Specific conductance (micromhos)	Specific conductance is a measure of the electrical conductivity of water: it varies with the amount of dissolved solids and is used to approximate the dissolved-solids content.	In the eastern and far northwestern parts of the United States most water has a specific conductance of less than 1,000 micromhos. In arid regions of western United States water with a specific conductance of more than 1,000 micromhos is not uncommon. Ocean water has a specific conductance of more than 50,000 micromhos.		1,660	8	1,660	18
pH (pH units)	pH values range from 0 to 14. Water with a pH of 7.0 is neutral. Water having a pH less than 7.0 is acid, and water having a pH greater than 7.0 is alkaline.	pH of ground water commonly ranges from 6.0 to 9.0. In surface water it commonly ranges from 6.0 to 8.0. Water influenced by acid mine drainage may have a pH about 2.0.	For most domestic and industrial uses, water having a pH between 6.0 and 10 generally causes no great problems. Water having a pH below the range may be corrosive.	9.8	6.0	10.5	5.0

Constituent	Source	Significance				
Color (color units)	Decaying vegetation; peat, lignite, and other plant remains; and industrial wastes cause color in water.	Ground water generally has little or no color. Surface-water swamps, where vegetation is abundant, may have color amounting to several hundred units on the cobalt-platinum scale.	Color due to suspended matter is generally removed during flocculation and filtration. Highly colored water is aesthetically objectionable for drinking water and is objectionable for many industrial processes, such as dyeing, brewing, and ice making.	115	0	24	0
Turbidity (parts per million)	Silts and clays from soil erosion, thermal turnover of lakes, and industrial wastes cause water to be turbid. Streams and lakes become turbid after intense rainstorms.	Ground water generally has little or no turbidity. In surface water the concentration may temporarily exceed 2,500 ppm.	Turbid water is aesthetically objectionable. Sediments causing turbidity may settle out and form films. Turbid water is objectionable for many industrial processes; turbidity is generally removed by sedimentation, clarification, or filtration.	2,170	0	13	0
Temperature (degrees Fahrenheit)	Surface-water temperatures approximate mean monthly air temperatures and ground-water temperatures approximate mean annual air temperatures. Shallow water is more sensitive to changes in air temperatures. Warmed industrial outflows raise stream temperature.	During winter, shallow streams and lakes may freeze. Temperatures of most streams are consistently less than 100°F. Thermal pollution has raised temperatures in some streams to more than 130°F.	Warm drinking water is objectionable. At times, warm water is advantageous for some water-treatment and industrial processes. However, for industrial cooling water, generally the higher the water temperature, the larger the amount of water required.				

Figure 2 shows the hardness of untreated water sources for these largest cities. For cities that obtain their water from more than one source, the hardness is weighted according to the population served from each source. Many of the hardness calculations are based on yearly averages; others are based on samples collected to represent an average hardness of water. Some cities are not included here because of the lack of data.

For ordinary household uses and for many industrial purposes, "soft" water (hardness 0—60 ppm) requires no softening. However, softening is required by a few industries and the operation of some steam boilers at pressures in excess of 200 pounds per square inch. Twenty-nine cities, serving more than 21 million people, have "soft" raw water and do not soften their supply. Water having a low hardness may become corrosive; therefore, some of these cities add lime to raise the pH and thus slightly increase the hardness.

Specific Conductance

Specific conductance is a convenient rapid determination used to estimate the amount of dissolved solids in water. It is a measure of the ability of water to transmit a small electrical current. The more dissolved solids in water that can transmit electricity, the greater the specific conductance of the water. Commonly, the dissolved solids (in parts per million) is about 65 percent of the specific conductance (in micromhos). This relationship is not constant from stream to stream or from well to well, and it may even vary in the same sources with changes in the composition of the water.

For highly mineralized water and highly colored water, the dissolved solids is more than 65 percent of the conductivity; for water containing large amounts of acid, caustic soda, or sodium chloride, the dissolved solids is less than 65 percent of the conductivity.

pH

Water that is neither acidic nor basic (alkaline) is called a neutral water and has a pH of 7.0. The pH of water solutions can range from 0 to 14 and can be increased or lowered by the introduction of chemicals. Strong acids—such as sulfuric, hydrochloric, or nitric acid—added to a neutral water can reduce the pH to as low as 0. Weak acids, like carbonic acid, added to water also lower the pH, although not as effectively as strong acids. Conversely, strong bases—like sodium hydroxide (caustic soda)—can increase the pH of water to as much as 14; weak bases do not increase the pH of water as effectively as strong bases. Salts formed by the reaction of strong acids and strong bases generally have little effect upon the pH of neutral water. Salts of a strong acid and of a weak base—such as iron sulfate—when added to a neutral water lower the pH, and salts of a weak acid and a strong base—such as sodium carbonate—increase the pH of a neutral water.

Geologic terrane and environment influence the pH of streams, lakes, and underground water. Most rocks in contact with water are not very soluble, and most streams and underground water have only small amounts of dissolved solids. In these dilute solutions, the introduction or the loss of small amounts of chemicals can radically alter the pH. For example, when well water having a pH

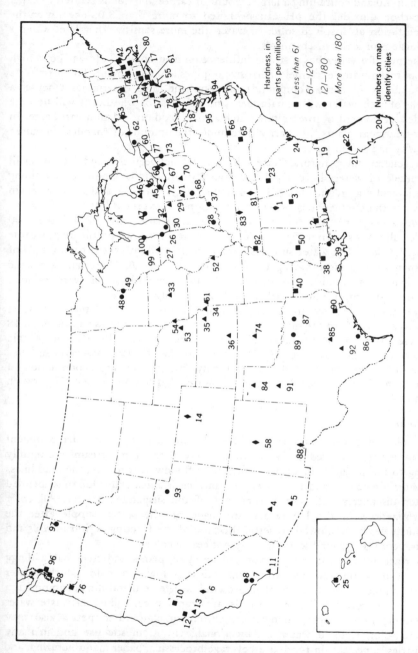

FIGURE 2. Hardness of untreated public water supplies of the 100 largest cities in the United States, 1962, Several cities are not shown because of insufficient data. (Average weighted by population served.)

Hardness, in parts per million

■ Less than 61

◆ 61–120

● 121–180

▲ More than 180

Numbers on map identify cities

less than 5.0 and containing a large amount of carbon dioxide is aerated to expel the carbon dioxide, the pH can be raised to more than 8.0. The lower the concentration of dissolved solids in water, the more sensitive the pH of water is to additions or losses of chemicals.

The pH of a water has a strong influence on its usability. A low pH or a high pH can make water extremely corrosive to pipes and equipment. The pH affects the solubility of some compounds in water and thus determines whether a sample of well water will remain clear and colorless or whether it will become clouded or colored by precipitates such as iron oxide or calcium carbonate. In water-treatment plants, pH partly determines the amount of chemicals required to clarify and soften water.

In general, most natural waters have a pH between 5.0 and 8.0. A small percentage of waters have a pH less than 5.0. Acid mine drainage containing sulfuric acid may reduce the pH of streams to less than 2.0, and some waters in contact with extremely basic rocks can have a pH in excess of 9.0.

The average pH of the raw-water resources used by 98 of the 100 municipalities is shown in figure 3. Two cities are not shown on the map because of lack of data. The water supplies have been grouped into those having a pH of less than 7.0 and those having a pH between 7.0 and 9.0. These data are based on calculations submitted by officials of city waterworks or on water samples that are representative of the pH for the water supply.

The water used by these largest cities is obtained from the best water resources available to the municipal water departments. As shown in figure 3, 18 of these cities, serving a total population of more than 16 million people, obtain raw water that has an average pH of less than 7.0; the pH of all raw-water supplies in these cities was between 5.8 and 7.0. Eighty cities, serving a total population of more than 42 million people, obtain raw water that has an average pH between 7.0 and 9.0.

Color

The color of streams and lakes is caused principally by suspended sediment and by matter dissolved in water. Immediately after a rain, streams are muddy owing to the sediment in suspension. As the floodwaters recede, the muddiness of water disappears, and the water becomes clear. Most color due to suspended matter disappears with the settling out of the suspended matter. All color determinations in the laboratory are made on the water sample after the sediment has been allowed to settle. Because of the filtering action of soils and rocks, very little ground water has any noticeable color.

Surface water containing living and decaying plants and trees has a dingy tinge. During the summer when streamflow is low, the color of the water becomes accentuated because plant growth is accelerated and the decomposition of decaying vegetable litter proceeds at a rapid pace. Industrial waste water containing iron, copper, manganese, chromium, and other metals also may impart color. Colored water is objectionable for domestic use and in many industries, especially in food and beverage processing, paper manufacturing, and dyeing industries.

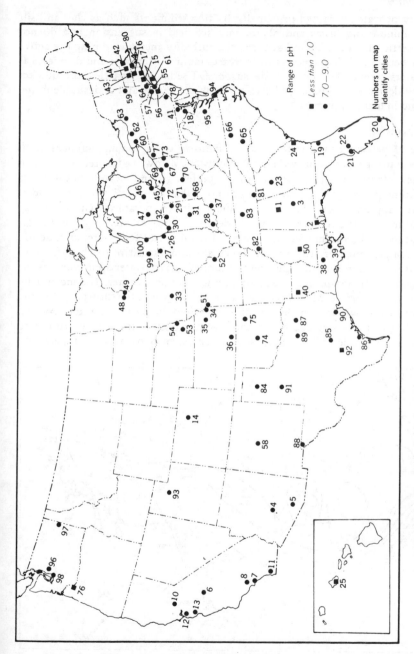

FIGURE 3. pH of untreated public water supplies of the 100 largest cities in the United States, 1962. Two cities are not shown because of insufficient data. (Average weighted by population served.)

Range of pH

■ *Less than 7.0*

● *7.0–9.0*

Numbers on map identify cities

Turbidity

Turbidity of water is caused principally by fine sediments such as clay and silt and by minute organisms and plants that are held in suspension and do not rapidly settle out. In lakes and streams the turbidity increases during the active growing period and, like color, also increases rapidly after rains and decreases as floodwaters recede. The heavier the suspended sediment particles, the quicker turbidity decreases. Turbidity, like color, is objectionable and undesirable in the home and in many industries.

Temperature

Because 95 percent of the water used by industry is for cooling, temperature is an important property of water. A consistently low water temperature is desirable. Many industrial water users prefer ground water because its temperature generally does not change more than $3°-4°$ F per year, and it generally approximates the mean annual air temperature. Ground-water temperature tends to increase with depth; below 60 feet, ground-water temperature increases only about $1°$ F for each $60-100$ feet increase in depth.

The temperatures of streams and lakes are more sensitive to changes in air temperature. The mean monthly temperature of surface water approximates the mean monthly air temperature, except during freezing weather. The mean daily temperature of surface water increases at a slower rate in the spring months and decreases at a slower rate in the autumn than does the mean daily temperature. The shallower the water depth, the more sensitive the water temperature is to changes in air temperature. Figure 4 is a general map of stream temperatures

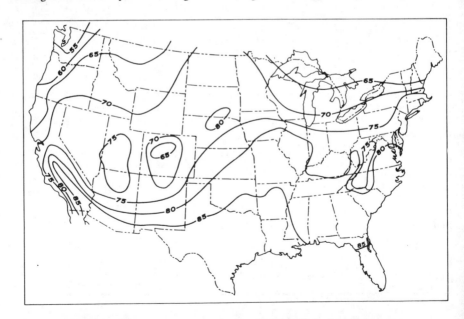

FIGURE 4. Stream temperatures, in degrees Fahrenheit, during the summer.

compiled from 467 maximum monthly mean temperature readings (U.S. Geological Survey, 1962).

References

1. Aultman, W. W. 1957. Softening of municipal water supplies. Water and Sewage Works. 104:327-334.
2. Baker, M. N. 1948. The quest for pure water. New York; Am. Water Works Assoc., 527 p.
3. DeBoer, L.M., and T.E. Larson. 1961. Water hardness and domestic use of detergents: J. Am. Water Works Assoc. 53:809-822.
4. McNeil, D. R. 1957, The fight for fluoridation. New York; Oxford Univ. Press, 241 p.
5. Moore, E. W. 1940. Progress report of the committee on quality tolerances of water for industrial uses. New England Water Works Assoc. Jour. 54:263.
6. Soap and Detergent Assoc. 1962. Synthetic detergents in perspective—their relationship to sewage disposal and safe water supplies. New York: Soap and Detergent Assoc., 39 p.
7. U.S. Department of Health, Education, and Welfare, Public Health Service. 1962. Fluoride drinking waters. Public Health Service Pub. 825, 636 p.
8. U.S. Department of Health, Education, and Welfare, Public Health Service. 1962b. Drinking water standards. Public Health Service Pub. 956, 61 p.

Part Two

Sources of Surface-Water Pollution

C oal-mining activities, especially in Appalachia, have for decades grossly affected the chemical quality of water in streams and lakes and underground. Similar mining techniques in the extensive lignite areas in the Great Plains have not caused similar water-quality problems. The major reason for this difference is due to the mineral makeup of the coal-bearing rocks. Unlike those in the Great Plains, the strata in Appalachia contain rather large amounts of iron sulfide minerals that are rapidly weathered. When leached by rain from unsaturated ground, these soluble substances form a water that is acid (low pH) and highly mineralized with large concentrations of sulfate and iron. This water is highly corrosive. The regional magnitude of the coal mine drainage problem is described by J. E. Biesecker and J. R. George.

Lake Erie has often been called a "dead lake" mainly because of abundant algal problems. Notwithstanding, this lake continues to produce more fish than all of the other Great Lakes combined. Lake Erie is not dying; on the contrary, it is too productive. Huge quantities of agricultural, municipal, and industrial wastes pour into the lake each day. Many of these wastes provide an over-abundance of nutrients permitting extensive algae blooms. The source and quantity of many of these wastes are succinctly illustrated in *The Lake Erie*

Photo opposite courtesy of Henry H. Valiukas, MECCA, St. Paul, Minn.

Report, a single chapter of which is included in these readings. B. L. Schmidt gives an account of the types of pollutants derived from farming, namely sediment and agricultural chemicals. With the possible exception of nitrate, these substances have their most profound effect on surface waters.

Most of our lakes and streams are much less contaminated than Lake Erie or the water courses in Appalachia. Moreover, it is not the intention of the following reports to suggest that all of our surface supplies are grossly polluted, for indeed they are not. There is cause for concern, however, and timely citizen-supported action is needed to provide comprehensive controls on waste disposal.

Although Part Two deals exclusively with surface water, this division is arbitrary. From a hydrologic point of view, it is unrealistic to separate surface water from ground water. We find, for example, that streams flow during rainless periods because of the ground water that seeps into the stream channel. Thus, if a stream-side aquifer becomes contaminated, this water will eventually find its way into a water course causing a degradation in quality. Conversely, contaminated surface water may be induced, either naturally or artificially, into the ground and contaminate the ground water. Examples of such situations are found in Parts Four and Five.

Stream Quality in Appalachia as Related to Coal-Mine Drainage, 1965

J. E. Biesecker* and J. R. George**

Introduction

Extensive coal mining in the Appalachian region for several decades has measurably influenced stream quality throughout the area. The deterioration of streams that receive coal-mine drainage has seriously limited the industrial and domestic uses of these waters. This undesirable alteration of natural stream quality has placed economic restrictions on many downstream water users.

Highly detailed individual studies and some broad statewide studies of the mine-drainage problem have varied greatly in technical approach as well as analytical methodology. However, most attempts to understand the problem and to define the extent of stream-quality damage were not designed to measure the relative significance of mine drainage on the water resources of the entire region. Public awareness of the problem balanced by technical concern over water pollution warrants a broad look at water pollution from coal-mine drainage throughout the entire 11-State area known as Appalachia.

Purpose and Scope

To evaluate the significance of this water-pollution problem in Appalachia, to update stream-quality data, and to provide technical continuity in collection of these data, the Geological Survey acquired extensive stream-flow and water-quality information in May 1965. This report summarizes the results of the first major regional reconnaissance, describes some basic water-quality characteristics of streams in the area, discusses the observed effects of mine drainage upon stream quality, and delineates areas where stream pollution by coal-mine drainage was most severe during the period of study.

The authors wish to make it clear that this report may present only limited new evidence of stream pollution by mine drainage to those interested in any

Reprinted from U.S. *Geological Survey Circular 526*, 1966. Reprinted version is slightly modified owing to deletion of original basic data table and plate.
*Water Resources Division, U.S. Geological Survey, Denver, Colorado.
**Water Resources Division, U.S. Geological Survey, Atlanta, Georgia.

specific part of Appalachia. The reconnaissance study is intended primarily to offer a means of assessing the magnitude of this water problem throughout the entire region. This report provides a foundation of data to guide future regional studies. It also should assist in the selection of areas that require special, more detailed attention.

Acknowledgments

The authors are grateful to the many Water Resources Division offices which participated in the collection and appraisal of data presented in this report. We also wish to express thanks to the water resources agencies of the States in the Appalachia region for their cooperative support.

Coal Mining and Mine Water

The Appalachia region, as defined in Public Law 89—4 (1965), extends over parts of an 11-State area from Pennsylvania to Alabama (fig. 1). A small area in New York was added to Appalachia after enactment of Public Law 89—4. However, this report covers only that part of Appalachia south of the New York-Pennsylvania boundary. Coal deposits occur in approximately 50,000 square miles of the region and are in 9 of the 11 States. In many areas of Appalachia, coal has been intensively mined for more than 100 years. Records of production in some of these areas are available since post-Civil War years. Table 1 shows the amount of coal mined in Appalachia for the period 1923-63. It is noteworthy that 72 percent of the coal was mined in two northern States— Pennsylvania and West Virginia. Kentucky, Ohio, Virginia, and Alabama also produced significant amounts. . . .

Underground mining has produced most of the coal in Appalachia. However, recent data reflect a major trend to produce coal by strip mining. Since 1940 the amount of coal produced in the United States by strip mining has increased from 9 to 34 percent (U.S. Bureau of Mines, 1963). In 1963 strip mining outproduced other methods of mining in Ohio and Maryland (table 2).

While the quantitative significance of various types of coal mining upon water

TABLE 1. Total coal production in Appalachia, 1923-63.

State	Total coal produced,[1] in thousands of tons	Percent of total
Alabama	579,000	3.7
Georgia	1,000	.0
Kentucky	1,703,700	10.8
Maryland	61,600	.4
Ohio	1,097,200	6.9
Pennsylvania	6,294,400	39.7
Tennessee	230,700	1.4
Virginia	689,700	4.4
West Virginia	5,180,900	32.7

[1]U.S. Bureau of Mines (1963).

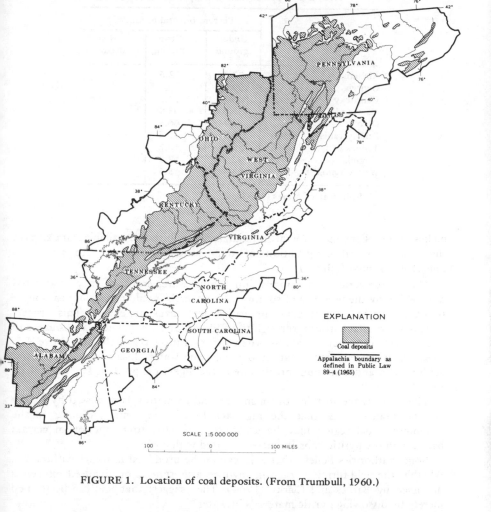

FIGURE 1. Location of coal deposits. (From Trumbull, 1960.)

pollution is not known, the problem, process, and products of mine drainage are similar for all types of mining. The problem begins with the physical process of unearthing coal which exposes pyritic materials (FeS_2), commonly associated with coals, to water and air. Braley (1954), Krickovic (1965), and Stumm (1965) state that pyrite reacts with oxygen and water to form ferrous sulfate ($FeSO_4$) and sulfuric acid (H_2SO_4). These chemical processes, whether within mines or waste piles, usually increase the concentrations of certain dissolved solids in the mine water. These index parameters include iron, sulfate, non-carbonate hardness, and total dissolved solids. Free mineral acidity, low pH values, and excessive concentrations of manganese and aluminum also are common characteristics of coal-mine waters. The sulfate in the reaction products

TABLE 2. Type of mining in Appalachia, 1963.[1]

State	Percent of total production		
	Under-ground	Strip mining	River dredging
Alabama	77.4	22.6	
Georgia	100.0		
Kentucky	66.7	33.3	
Maryland	36.6	63.4	
Ohio	33.7	66.3	
Pennsylvania	59.9	35.6	4.5
Tennessee	59.3	40.7	
Virginia	92.5	7.5	
West Virginia	94.4	5.6	

[1]U.S. Bureau of Mines (1963).

makes an excellent indicator of mine-drainage pollution. Available data suggest that the chemical composition of mine waters throughout Appalachia is remarkably similar (table 3).

The concentration and composition of mine water, however, may be affected measurably by the presence of soluble rock minerals including calcium carbonate ($CaCO_3$), which in sufficient quantities neutralizes mine acid. This process increases the total hardness through the addition of calcium and magnesium, and can increase carbonate hardness when neutralization raises the pH above 4.5. Even when partial neutralization occurs, mine waters lose some free mineral acidity. Iron and aluminum precipitate at the higher pH produced by neutralization.

The significance of microorganisms in acid formation is discussed by Braley (1954). Braley states that the high acidity of many mine effluents in the bituminous coal region may be attributed, in part, to the action of certain bacteria on the pyritic constituents associated with coal.

Some authorities believe that acid cannot be produced in mines without air. Whether the acid-forming reaction involves atmospheric or dissolved oxygen is discussed by Barnes and Clarke (1964). They suggest that acid can be formed merely by dissolving pyritic materials in water.

Mine Drainage and Stream Quality

The delivery of mine water to the surface drainage system is a critical factor in controlling the extent of stream pollution by coal-mine drainage. Relatively continuous delivery of water to a stream from active and abandoned coal mines creates continuous pollution of the stream. This type of stream-quality damage is of great concern to industrial and domestic users who must maintain extensive treatment facilities to obtain a usable supply of water.

Occasional flushing of mines by excessive precipitation produces temporary, but often more dramatic, stream damage. Mine flushing delivers a large volume of water to a stream for a short time. When this mine effluent is carried

TABLE 3. Chemical composition of typical mine waters in Appalachia. [Results given in parts per million except as indicated]

Mine name	Location	Date	Silica (SiO_2)	Aluminum (Al)	Iron (Fe)	Manganese (Mn)	Calcium (Ca)	Magnesium (Mg)	Sodium (Na)	Potassium (K)	Bicarbonate (HCO_3)	Sulfate (SO_4)	Chloride (Cl)	Fluoride	Nitrate (NO_3)	Dissolved solids (residue on evaporation at 180°C)	Hardness as $CaCO_3$ — Calcium, magnesium	Noncarbonate	Total acidity as H_2SO_4	Specific conductance (micromhos at 25°C)	pH	Color
PENNSYLVANIA																						
Delaware River basin																						
Newkirk mine	Near Tamaqua	7-28-65	25	40	5.5	10	60	49	1.5	1.4	0	672	1.2	0.2	0.8	1,060	351	351	382	1,590	2.80	3
Eagle Hill 2	Near Port Carbon	7-28-65	9.5	---	2.8	2.2	84	51	4.3	1.0	84	344	.8	.2	.2	571	420	351	---	765	7.8	3
Susquehanna River basin																						
Middle Creek mine.	Near Tremont	6-25-65	16	5.6	9.2	4.9	61	48	2.0	1.5	0	427	3.2	0.3	0.4	671	350	350	108	1,080	2.95	1
Glenwhite Run mine 6.	Near Altoona	6-16-65	---	---	10	---	62	---	---	---	0	900	---	---	---	---	680	680	717	1,740	3.1	--
Allegheny River basin																						
Toby Creek mine 2.	Near Brandy Camp.	6-17-55	---	---	164	---	214	---	---	---	0	1,730	---	---	---	---	1,560	1,560	1,170	3,080	2.7	--
KENTUCKY																						
Big Sandy River basin																						
Cane Branch mine 1.	Near Wayland	10-9-58	34	14	18	---	152	28	38	5.6	0	832	2.5	1.9	5.0	1,200	494	494	236	1,870	2.80	--
Ohio River basin																						
Yellow Creek mines.	At Sassafras	1-29-58	89	---	119	9.3	---	---	---	---	0	1,240	---	---	---	1,700	374	374	868	2,180	2.50	--
WEST VIRGINIA																						
Monongahela River basin																						
Norton mine 1	Near Norton	7-15-65	38	48	114	5.1	---	---	---	---	0	1,150	---	1.4	5.4	1,820	630	630	587	2,330	2.70	--
Browns Creek mine A-2.	Near Mount Clare.	9-12-63	3.3	16	217	4.3	---	---	79	8.7	0	1,960	---	---	---	---	1,130	1,130	490	3,210	2.75	--

downstream to points that normally are not affected by critical levels of pollutants, a fish kill may occur. The West Branch Susquehanna River in Pennsylvania, for example, experienced 20 major fish kills between 1948-62 (Corps of Engineers, 1962) because of the downstream transport of mine effluents by highly localized rains in the mining region.

The cumulative influence of both continuous and occasional stream pollution on fish habitat has been considered by the U.S. Bureau of Sport Fisheries and Wildlife (Kinney 1964). Kinney reports that Pennsylvania and West Virginia contain over two-thirds of the stream mileage that is adversely affected by coal-mine drainage in Appalachia. There is a striking relationship between Kinney's data and coal production by each State in Appalachia (fig. 2).

Although both continuous mine drainage and mine flushouts are of great

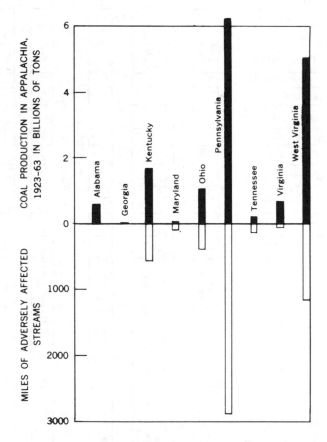

FIGURE 2. Comparison of coal production (data from U.S. Bureau of Mines, 1964) to miles of affected streams (data from U.S. Dept. Int., Div. of Fishery Management Services, E. C. Kinney, 1964).

public concern, this broad reconnaissance study defines only a part of the continuous effects. In particular, the May field studies measured the influence of mine drainage on stream quality during near-median flow conditions when streams contained fairly dilute waters. Most continuous mine-drainage pollution problems observed in May should be more serious during the June to November low-flow period when stream waters normally are more concentrated.

Wherever stream-water quality is affected seriously by coal-mine drainage, many economic limitations are placed on the value of that water for recreational, industrial, and municipal uses. An .abundance of mine-drainage constituents increases water-treatment costs and necessitates more frequent replacement of water-treatment facilities. River structures and navigation equipment often need special protection from corrosion by mine drainage. Deposits of sediment create an unattractive environment and render streams and lakes that receive mine discharge unfit for fishing, swimming, and other recreational uses.

Table 4 summarizes common water-use limitations of mine water. These include unpleasant taste, staining, increased corrosiveness, formation of insoluble precipitates, and unpleasant diuretic effects and are caused by excessive concentrations of the mine-drainage index parameters.

Stream-Quality Observations

The Field Reconnaissance

Available geologic, hydrologic, and coal-mining data were used to select sampling sites for the field reconnaissance. Because broad definition was a primary goal, many water-quality measurements in the 11-State region were made for streams draining an area greater than 100 square miles. In areas of coal mining, sampling sites were selected for streams known or suspected to be influenced by mine drainage, so that the relative stream quality could be assessed. Several additional sites were selected to represent the water quality of streams not affected by mine drainage.

The field reconnaissance in the late spring of 1965 was intended to define general water quality for near-median streamflow conditions. Unregulated streamflow during the study was in the 45-65 percentile range and provided comparative data on the influence of mine drainage on streams in the entire region. Streamflow was generally steady and, therefore, water-quality results were not complicated by the effects of direct runoff from rains. Consequently, most of the analyses provide areawide data on near-average water quality.

During the intensive 9-day study period in May 1965, 11 two-man teams of hydrologists and chemists visited 318 stream sites from northeastern Pennsylvania to central Alabama, an area of more than 160,000 square miles. Field measurements included water discharge, pH, specific conductance, water temperature, dissolved-oxygen concentration, and acidity. Water samples for more detailed analyses were also collected for delivery to U.S. Geological Survey laboratories in the region.

TABLE 4. Use limitations of water-quality parameters typical of coal-mine drainage.

Constituent	Objectionable features of excessive concentration	Recommended limiting concentration for indicated use (ppm)[1]						
		Public water supply[2]	Cooling water	Food processing	Pulp and paper making	Plastics manufacturing	Boilers	Textile manufacturing
Sulfate	Diuretic effect, bitter taste.	250	---	20–250	---	---	---	100
Hardness as CaCO₃.	Boiler scale, produces insoluble "curd" when it reacts with soap.	---	50	10–400	100–200	---	2–80	0–50
Dissolved solids.	Diuretic effect, unpleasant taste.	500	---	850	200–500	200	50–3,000	---
Iron	Unpleasant taste, stains porcelain and linen.	.3	.5	.2	.1–1.0	---	---	.1–1.0
Manganese	Unpleasant taste, stains porcelain and linen.	.05	.2–0.5	.2	.05– .5	.02	---	.1–1.0
Aluminum	Boiler scale.	---	---	---	---	---	0–3	---
Suspended solids.[3]	Clogs treatment facilities and water courses.	5	50	1–10	10–100	---	0–10	.3–25
pH[4]	Increases corrosiveness.	---	---	7.5	---	---	8.0–9.6	---

[1]California Water Quality Control Board (1963).
[2]U.S. Public Health Service (1962).
[3]Turbidity, as silica, in parts per million.
[4]Value not to be less than limits shown.

Basic Quality of Streams in Appalachia

Streams in the coal region that are unaffected by mine drainage are of excellent quality. These streams contain very dilute alkaline water, with calcium and bicarbonate the dominant dissolved constituents. During the study period, the bicarbonate content of unpolluted streams in the coal-mining region generally was less than 50 ppm. Unaffected streams adjacent to the coal region contained bicarbonate concentrations from 50 ppm to more than 200 ppm.

The alkalinity of streams in the southeast edge of Appalachia was generally lower than that observed for other streams within the region. Here streams draining the crystalline-rock terrain of the Piedmont Province contain among the lowest solute content of streams in the Eastern United States (Rainwater, 1962).

Unusually low concentrations of bicarbonate for unaffected streams in the coal-mining region demonstrate the relative inability of most of these streams to neutralize acid-mine water which enters the drainage system. When acid drainage from coal mines reacts with the low natural alkalinity of most streams in the coal region, the result is a large number of seriously affected streams carrying free mineral acidity.

While the May 1965 reconnaissance suggests that unaffected streams in the coal region contain relatively little neutralizing capacity, the bicarbonate alkalinity in some streams affected by mine drainage in parts of Pennsylvania, Ohio, West Virginia, Kentucky, and Virginia, indicates that extensive neutralization takes place within the coal region. Figure 3 illustrates a general area in the coal region where affected streams contain high concentrations (50-200 ppm) of bicarbonate. These high alkalinities may be produced by neutralization from small, highly alkaline tributaries that were not sampled during this reconnaissance. Neutralization also may occur in the mines by contact of water with adjacent calcareous rocks (or by mixture of alkaline water associated with these strata). Scattered evidence of the existence of alkaline mine waters add credibility to the second choice, but it may be a combination of these conditions that produces generally high stream alkalinity in the area noted in figure 3.

Hardness is another water-quality index parameter in which major changes usually occur when mine waters are added to natural streamflow. The ranges in concentrations, zonal boundaries, and related criteria used to describe alkalinity of both unaffected and affected streams are similar for total-hardness data collected during the reconnaissance. Where mine drainage has not influenced stream quality in the coal region, total hardness was nearly always less than 50 ppm. In the areas immediately adjacent to the coal region, total hardness ranged from 50 ppm to 300 ppm or more. Again, streams draining the Piedmont province of southeastern Appalachia were most dilute, the hardness values ranging generally from 10 to 20 ppm.

Salty water brought to the earth's surface while developing oil and gas wells often affects stream quality. The May 1965 data indicate that only a few major streams in Ohio, Kentucky, and Pennsylvania contained concentrations of chlorides in excess of 100 ppm, and the concentrations exceeded U.S. Public Health Service (1962) "Drinking Water Standards" of 250 ppm at only two sites.

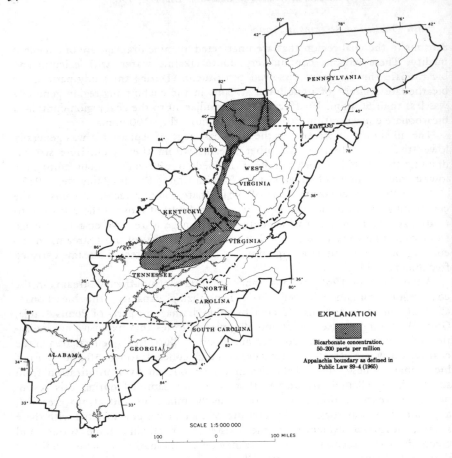

FIGURE 3. Areas in the Appalachian coal region where streams affected by mine drainage contained relatively high (50-200 ppm) concentrations of bicarbonate, May 1965.

The chloride data indicate that although some brine pollution does exist in Appalachia, it is not a major problem on large streams during median flow.

The nitrate and phosphate content of stream water is considered a secondary indicator of pollution from untreated or treated domestic wastes as well as from some industrial wastes. In Appalachia, observed concentrations of nitrates and phosphates were low during the May 1965 reconnaissance. Concentrations of both constituents were well below recommended limits for public water supplies and also were acceptable for most industrial uses of water. The lack of these constituents in water, in fact, suggests a deficiency of some key nutrients that fertilize aquatic plants. This deficiency may provide a poorer environment for many types of aquatic insects and fish which, in turn, can exert some limitation on recreational development of the water.

Dissolved-oxygen concentration may also serve as an indicator of pollution by domestic and industrial waste. Although observed dissolved-oxygen values represent only an instantaneous evaluation of a complex and dynamic system of stream deoxygenation and reaeration, data collected during the reconnaissance offer means for a limited appraisal of stream conditions. In Appalachia, most observed dissolved-oxygen concentrations were above the suggested value of 5.0 ppm (California Water Quality Control Board, 1963, p. 181), necessary for a favorable environment for fish and other aquatic life. The dissolved-oxygen concentration was less than 5.0 ppm at only 10 of 318 locations.

Effects of Mine Drainage on Stream Quality

The presence of free mineral acidity in a stream is the most serious evidence of water-quality damage by mine drainage. The May 1965 data clearly demonstrate that mine drainage damages the chemical quality of streams more severely in the northern one-third of Appalachia than in the rest of Appalachia. Free mineral acidity occurs in rivers as large as West Branch Susquehanna River, Kiskiminetas River, Casselman River, North Branch Potomac River, Monongahela River, and Raccoon Creek.

The abundance of acid mine waters in northern Appalachia may be due to several factors or to a combination of these factors. There is more coal mined in the north than in the south. This implies more extensive exposure of sulfuritic material to an acid-producing environment. Also, the amount of sulfuritic material exposed for each ton of coal mined in the north may be greater than in the south.

Yearly acid loads at several locations are reported in U.S. Geological Survey Hydrologic Investigations Atlas HA—198 (Schneider and others, 1965). The May 1965 data illustrate the immense magnitude of the mine-drainage problem in the West Branch Susquehanna River, Monongahela River, and Kiskiminetas River basins where the loads of acid per square mile are greater than those of other major basins in Appalachia.

The key index solute, sulfate, is used in this report to describe the influence of mine drainage on stream quality during median flow. Since observed concentrations of sulfate for unaffected streams draining the coal region were low (less than 20 ppm) during the study period, concentrations of sulfate greater than 20 ppm are used to describe the measured effect of mine drainage on stream quality. The chemical quality of most major tributaries of the Susquehanna and Ohio Rivers that drain the Appalachian coal fields is affected to some extent by mine drainage. In the northern one-half of the coal region, only a few streams draining an unmined part of the Kanawha River basin are not influenced by mine drainage using sulfate concentration as an indicator of mine drainage. Farther south in parts of Tennessee, Georgia, and Alabama, scattered mining has little effect on the chemical quality of major streams in the area during median flow.

Figure 4 shows the north to south trend of sulfate content for streams affected by mine drainage. The median concentration of sulfate for affected streams in Pennsylvania and Ohio is 160 ppm, but only 45 ppm for streams in

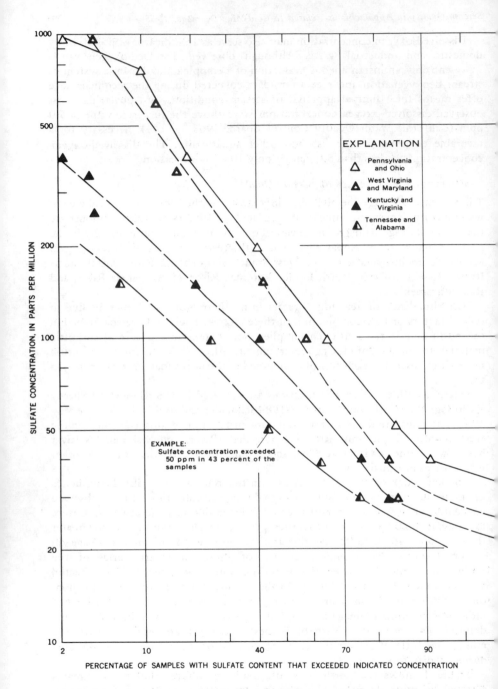

FIGURE 4. Effect of coal-mine drainage on sulfate content from northern to southern Appalachia, May 1965.

Tennessee and Alabama. This decrease in sulfate concentration provides further evidence of less intense mine-drainage problems in southern Appalachia.

The effect of mine drainage on the hardness of water is shown in figure 5. Note the greater percentage of samples in the hard and very hard class for mine-polluted waters. Median hardness was 130 ppm for affected sites, and only 30 ppm for unaffected sites.

With U.S. Public Health Service (1962) "Drinking Water Standards" as a guide for defining the limitations placed on stream use by mine drainage (table 4), it is apparent that mine drainage has seriously affected the utility of many streams in the region for domestic or municipal supply. Water quality at nearly 200 sites in the region did not meet recommended water standards. Table 5 describes the effects of mine-drainage index solutes on the potential use of these waters for municipal supplies. Several streams will not meet the water-quality criteria for many industrial uses of water.

Neutralization of Acid Streams

The mixture of alkaline streams with mine-drainage waters eventually neutralizes all acid streams in Appalachia. Even in the badly polluted upper Ohio River basin, the added flow from the Allegheny River and other more alkaline

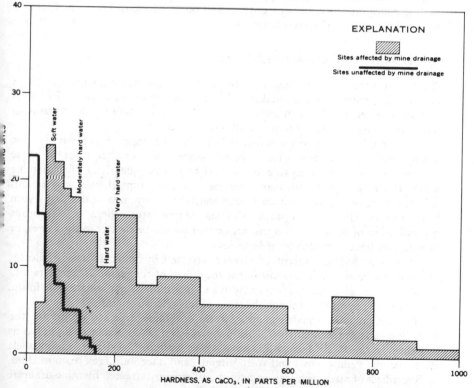

FIGURE 5. Effect of mine drainage on the hardness of water, May 1965.

TABLE 5. Effect of mine drainage on the potential use of
streams draining the coal region of Appalachia, May 1965.

Water-quality parameter	Percentage of sample sites where concentrations exceeded drinking water standards	
	Sites unaffected by coal-mine drainage	Sites affected by coal-mine drainage
Iron	6	35
Manganese	34	83
Sulfate	0	22

downstream tributaries ultimately produces water of fair quality (fig. 6). Observed pH increases from less than 4.7 in the Monongahela River to approximately 7.0 in the Ohio River at Stratton, Ohio. There is also a gradual increase in the ratio of bicarbonate to sulfate between these two sites. Stream hardness continues to increase downstream, but a greater part of the hardness is carbonate hardness.

Thus, while the problem of stream pollution by mine drainage is particularly serious in headwater areas of Appalachia near active and abandoned mines, the alkaline contribution of streams both in and out of the coal region measurably improves the quality of affected waters.

Conclusions and Recommendations

The May 1965 field reconnaissance discloses that the water quality at 194 of 318 sampling sites was measurably influenced by mine drainage. Thirty sites contained water with free mineral acidity. Nearly all major acid streams in Appalachia were in the northern one-third of the region.

The natural alkalinity of streams within the coal region was generally low, usually less than 50 ppm. The reconnaissance discloses that most of these streams are relatively incapable of neutralizing large quantities of acid mine water. However, high bicarbonate content was fairly typical of many streams affected by mine drainage within a comparatively narrow area of the coal region from eastern Ohio to western Virginia. Apparently, there is significant neutralization of mine water in this area either by alkaline water in the mines, or by small unaffected headwater tributaries.

Analysis of sulfate content of streams affected by mine drainage indicates that less sulfate occurs in streams in the south half of the region. These data also provide further evidence that less acid water is produced per square mile in the south than in the north, probably because of less intense mining.

More serious water-quality damage from coal-mine drainage occurs in: the West Branch Susquehanna River, Kiskiminetas River, and Casselman River basins in Pennsylvania; North Branch Potomac River basin in Maryland; Monongahela River basin in Pennsylvania and West Virginia; and Raccoon Creek basin in Ohio.

Regardless of size or degree of acidity, all streams affected by mine drainage are ultimately neutralized by the inflow of alkaline water.

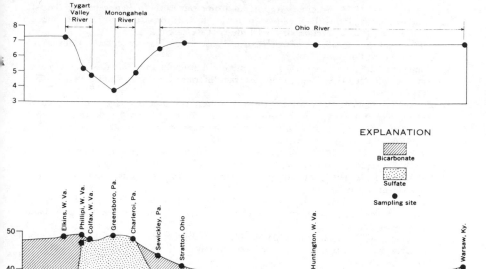

FIGURE 6. Changes in the composition of a mine-polluted stream during neutralization, Ohio River basin, May 1965.

Future regional studies should be designed to provide data for better definition of: (1) the significance of mine drainage upon stream quality during low-flow conditions, (2) types of hydrologic environment that produce sudden flushes of mine water to streams which normally contain relatively little mine water, and (3) the geologic and hydrologic factors contributing to high alkalinity of affected streams within isolated parts of the region.

References

1. American Public Health Association. 1960. Standard methods for the examination of water and wastewater. New York; Am. Public Health Assoc., p. 42-43.
2. Barnes, Ivan, and F. E. Clarke. 1964. Geochemistry of ground water in mine drainage problems: U.S. Department of the Interior, Geol. Survey Prof. Paper 473-A, p. A1-A6.

3. Braley, S. A. 1954. Summary report of Commonwealth of Pennsylvania, (Department of Health) Industrial Fellowship 1 to 7 incl. (From 20 August, 1946 through 31 December 1953). Mellon Inst. of Industrial Research, 279 p.

4. California Water Quality Control Board. 1963. Water quality criteria. Calif. Water Control Board Pub. 3-A, 548 p.

5. Kinney, E. C. 1964. Extent of acid mine pollution in the United States affecting fish and wildlife. U.S. Department of the Interior, Fish and Wildlife Service Circ. 191, 27 p.

6. Krickovic, Stephen. 1965. U.S. Bureau of Mines acid mine drainage control program and joint Interior-HEW Departments acid mine drainage control program. *In* Papers presented before symposium on acid mine drainage research, Mellon Institute, 20-21 May 1965, Pittsburgh, Pennsylvania, p. 111-126.

7. Mackenthun, K. M. 1965. Nitrogen and phosphorus in water. U.S. Department of Health, Education, and Welfare, Public Health Service Pub. 1305, 111 p.

8. Rainwater, F. H. 1962. Stream composition of the conterminous United States. U.S. Department of the Interior, Geol. Survey Hydrol. Inv. Atlas HA-61.

9. Rainwater, F. H., and L. L. Thatcher. 1960. Methods for collection and analysis of water samples. U.S. Department of the Interior, Geol. Survey Water Supply Paper 1454, p. 87-92.

10. Schneider, W. J., and others. 1965. Water resources of the Appalachian region, Pennsylvania to Alabama. U.S. Department of the Interior, Geol. Survey Hydrol. Inv. Atlas HA-198.

11. Stumm, Werner. 1965. Oxygenation of ferrous iron properties of aqueous iron as related to mine drainage pollution, *In* Papers presented before symposium on acid mine drainage research, Mellon Institute, 20-21 May 1965, Pittsburgh, Pennsylvania. p. 51-63.

12. Trumbull, James. 1960. Coal fields of the United States. U.S. Department of the Interior, Geol. Survey Mineral Inv. Resources Map MR-2651, sheet 1.

13. U.S. Army Corps of Engineers. 1962. Blanchard Reservoir, West Branch Susquehanna River, Pennsylvania. *In* Design memorandum 1, Hydrology and hydraulic analysis. U.S. Army Corps of Engineers.

14. U.S. Congress, House. 1944. Ohio River pollution control—Report of the United States Public Health Service. 78th Cong., 1st sess., House Doc. 266.

15. U.S. Congress, House. 1962. Acid mine drainage—A report prepared for the Committee on Public Works, House of Representatives, by Division of Water Supply and Pollution Control, Public Health Service, U.S. Department of Health, Education, and Welfare. 87th Cong., 2d sess., House Comm. Print 18.

16. U.S. Department of the Interior. 1965. Proposed outline of report required on strip and surface mining in the United States as authorized under Public Law 89-4, section 205 of the Appalachian Regional Development Act of 1965. U.S. Dept. Interior, Office of Assistant Secretary—Mineral Resources, p. 43-45.

17. U.S. Department of the Interior, Bureau of Mines. 1964. 1963 minerals yearbook. U.S. Bur. Mines, V. 2 Fuels, 533 p.

18. U.S. Department of Health, Education, and Welfare, Public Health Service. 1962. Drinking water standards. Public Health Service Pub. 956, 61 p.

19. Whetstone, G. W. 1963. Statement of water resources investigations in the Monongahela River basin. *in* Conference in the matter of pollution of the interstate waters of the Monongahela River and its tributaries. U.S. Dept. of Health, Education, and Welfare, Public Health Service, Conf. Proc., p. 31-147.

The Cause of Pollution in Lake Erie

Waste Sources

Municipal waste is the principal cause of pollution in Lake Erie and its tributaries. Another major cause of pollution, expecially in tributaries and harbors, is industrial wastes. These wastes consist not only of continuous, direct untreated discharges, but also combined sewer overflows and treatment plant effluents. Agricultural runoff is also a significant, but harder to define, source of pollution. Other sources of pollution are wastes from commercial and pleasure craft, spoil from harbor dredging, urban runoff, and shore erosion. All of these combined are now potentially disastrous to Lake Erie water quality.

Three geographical areas are primarily responsible for the present condition of Lake Erie (Table 1). These areas, in order of decreasing effect on the overall quality of Lake Erie water, are: (1) Detroit, Michigan and its surrounding municipalities, (2) the Cleveland-Cuyahoga River basin, and (3) the Maumee River basin. The Buffalo area has high waste inputs, but these wastes affect the Niagara River more than Lake Erie. Many other areas have problems which are primarily local; but cumulatively, they also have a profound effect on the general water quality.

The remarkably degrading effect which the Detroit, Cleveland, and Maumee areas have on Lake Erie can be shown by subtracting their discharges of almost any constituent from the total input of that constituent to the lake.

Municipal Wastes

Approximately 9 million people inhabit communities within the U.S. portion of the Lake Erie basin, discharging partially treated wastes directly into Lake Erie or into its tributaries. Nearly 2 million more people are served by septic tanks. Sixty-three municipal primary treatment[1] plants serve approximately 5 million

Reprinted from *Lake Erie Report, A Plan For Water Pollution Control*, Federal Water Pollution Control Administration, 1968, pp. 53-65. Several photographs have been deleted from the reprinted version.

[1] A settling process which removes about 35% of organic pollutants from sewage water.

TABLE 1. Percent waste contribution of
major source areas in Lake Erie basin.

	Detroit and Southeast Michigan	Cleveland Akron- Cuyahoga	Toledo- Maumee River
Phosphorus	40.0	18.6	15.3
Biochemical			
Oxygen Demand	60.3	11.0	15.5
Chloride	51.0	10.6	4.7

people, discharging 879 mgd of wastes; and 155 municipal secondary treatment[2] plants serve approximately 4 million people, discharging 591 mgd of wastes. Figure 1 shows the municipal treatment plant data by sub-areas. The ten largest sources of municipal waste based on strength of waste discharged are shown in Table 2. It is easily seen from the table that Detroit is the largest municipal polluter in the basin, and in fact contributes more wastes than all the other cities combined.

Almost the entire population of the Southeast Michigan area is in and around Detroit. The Detroit primary sewage treatment plant serves about 2.8 million people. *While 85 percent of the Southeastern Michigan population discharges wastes to sewers, the wastes from only 10 percent of the total population receive secondary treatment.* Southeast Michigan accounts for 58 percent of the entire municipal waste flow to the Lake Erie basin.

TABLE 2. Ten largest U.S. sources of municipal
waste in the Lake Erie drainage basin.

Location	% of total U.S. municipal waste BOD discharged	% of total U.S. municipal waste phosphorus load
Detroit, Michigan	64.4	46.5
Cleveland, Ohio	9.0	20.3
Toledo, Ohio	3.5	5.0
Wayne County, Michigan	3.1	2.8
Akron, Ohio	1.3	4.5
Euclid, Ohio	1.3	1.0
Lorain, Ohio	1.1	1.2
Sandusky, Ohio	1.0	0.6
Erie, Pennsylvania	0.9	2.2
Ft. Wayne, Indiana	0.5	2.6

[2] A biological process which removes up to 90% of organic pollutants from sewage water.

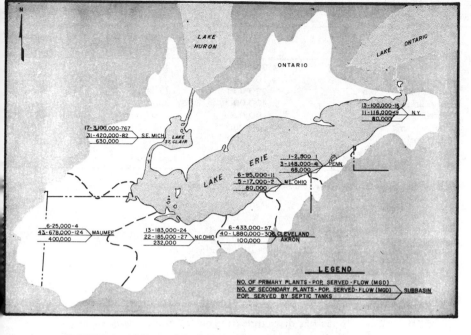

FIGURE 1. 1960 municipal sewage treatment plant data by subbasin.

About 79 percent of the Indiana Lake Erie basin population is served by secondary sewage treatment plants, and the rest, which is basically rural, is served by septic tanks.

About 64 percent of the total population in the Ohio portion of the Lake Erie basin is served by secondary treatment. About 16 percent of the population is not served by sewer systems. The Greater Cleveland-Akron area accounts for 25 percent of the municipal waste flow to the lake.

The wastes from almost 70 percent of the population in the Pennsylvania portion of the basin receive secondary treatment. The city of Erie, the largest city in Pennsylvania whose discharge reaches Lake Erie, provides secondary treatment.

In the New York area of the Lake Erie basin, the wastes from only 39 percent of the total population receive secondary treatment.

Biochemical oxygen demand (BOD) is the main municipal degradant discharged to tributary waters. The most harmful municipal contribution to the lake is in the form of nutrients—primarily nitrogen and phosphorus—although locally bacteria cause serious problems.

The total BOD discharged to municipal sewage treatment plants in the basin is equivalent to the raw sewage from a population of 9.4 million. After treatment, this is reduced to a load on the receiving waters equivalent to the raw sewage from a population of 4.7 million, or a treatment efficiency, basin-wide, of 50 percent.

TABLE 3. Municipal treatment plant waste discharges in population equivalents.

Subarea	Before treatment[1]	After treatment[1]	Treatment efficiency—%
Southeast Michigan	4,750,000	3,420,000	28
Maumee	1,100,00011	240,000	78
North Central Ohio	430,000	170,000	60
Cleveland-Akron	2,400,000	630,000	74
Northeast Ohio	100,000	60,000	40
Pennsylvania	360,000	42,000	88
New York	220,000	90,000	59
Total Lake Erie basin (U.S.)	9,360,000	4,652,000	50

[1]Equivalent to the oxygen demand of raw sewage from that number of people (0.17 pounds of BOD per capita per day).

Municipal waste treatment plant discharges and treatment plant efficiencies summarized by sub-basins are shown in Table 3.

Industrial Wastes

Industrial wastes are those spent process waters associated with industrial operations which are discharged separately and not in combination with municipal wastes. Industries discharge oxygen-consuming substances equivalent to the raw sewage discharge from a population of nearly 3 million. The major industries are listed below with pollution substances common to each:

Power	— Heat, Solids
Steel	— COD, Acids, Iron, Solids, Phenols, Oils, Color, Heat, Toxicants
Chemical	— COD, Solids, Organics, pH, Toxicants, Color
Oil	— Oil, COD, Solids, Phenols
Paper	— BOD, Solids, Color, Coliform
Rubber	— Organics, Oil, COD, Solids
Plating	— Machinery, Manufacturing — Cyanide, Chrome, Cadmium, Copper, and other heavy metals, pH, Solids
Food Processing	— BOD, Oil, Solids, Color

Lake Erie and its tributaries receive industrial wastes from 360 known sources. A summary of these, by states, is given in Table 4, along with their treatment adequacy. *Slightly more than 50 percent of the industries are classified as having inadequate treatment facilities.*

Industries are responsible for 87 percent of the total waste flow discharged to Lake Erie and its tributaries. The total waste volume from industry equals 9,600 mgd with electric power production accounting for 72 percent and steel production accounting for 19 percent of the flow. Figure 2 shows the industrial waste flow data by subareas.

The 20 largest industrial water users based upon volume of waste discharged exclusive of electric power production are shown in Table 5. It is easily seen

TABLE 4. Industrial waste source classification.

State	Treatment	
	Adequate	Inadequate
Ohio	119	92
Indiana	9	6
Michigan	24	68
Pennsylvania	15	5
New York	4	18
Total	171	189

from the table that steel, chemical, oil, and paper industries predominate in the basin. These 20 industries discharge 86 percent of the total industrial waste water if the electric power industry is excluded.

About 47 percent of the total industrial waste discharges in the basin flows directly to the lake or to lake-affected portions of the tributaries; another 44 percent is discharged to the Detroit and St. Clair Rivers.

Industries discharge millions of pounds of dissolved solids daily to Lake Erie; for example, they discharge 11 million pounds of chlorides daily and a similar quantity of sulfates. The chloride input is expected to double by 1990 unless restrictions are placed upon inputs.

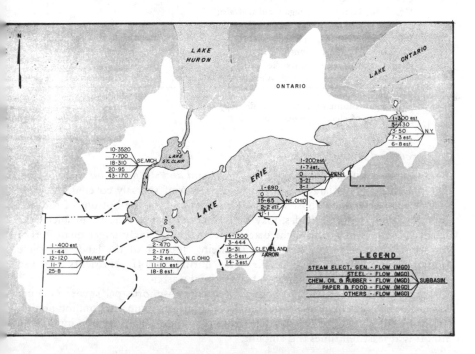

FIGURE 2. 1960 industrial discharges by subbasin.

TABLE 5. Twenty largest U.S. producers of industrial waste
water[1] in the Lake Erie drainage basin.

Name and location	% of total industrial waste discharge[2]
Ford; Dearborn and Monroe, Mich.	19.7
Republic Steel; Lorain and Cleveland, Ohio and Buffalo, N.Y.	14.9
Bethlehem Steel; Lackawanna, N.Y.	13.0
Great Lakes Steel; Ecorse and River Rouge, Michigan	8.7
Jones & Laughlin Steel; Cleveland, Ohio	4.8
Wyandotte Chemical; Wyandotte, Mich.	4.1
Pennsalt Chemical; Riverview, Mich.	3.6
Gulf Oil; Toledo, Ohio	2.5
McLouth Steel; Trenton and Gibraltar, Michigan	2.4
Allied Chemical; Detroit, Michigan and Buffalo, New York	1.7
Interlake Steel; Toledo, Ohio	1.6
Scott Paper; Detroit, Michigan	1.6
Standard Oil; Toledo and Lima, Ohio	1.5
Midland Ross; Painesville, Ohio	1.1
U.S. Steel; Cleveland and Lorain, Ohio	0.9
Mobil Oil; Trenton, Michigan and Buffalo, New York	0.9
Hammermill Paper Co.; Erie, Penn.	0.7
Monsanto Chemical; Trenton, Mich.	0.7
Diamond Shamrock; Painesville, Ohio	0.6
Consolidated Paper; Monroe, Mich.	0.5

[1] Based on volume of waste water discharged.
[2] Exclusive of electric power production.

Agricultural Runoff

Agricultural runoff is a major source of nutrient and silt pollution to Lake Erie.
The pollution results largely from surface erosion of sparsely covered, intensely
cultivated, fine-grained soils. While silt covers much of the bottom of the lake
and this, along with pesticides, may be influential in the fisheries problem, the
nutrient input is of greater immediate concern. Increasingly larger quantities of
nitrogen and phosphorus fertilizers are being applied to the land, and these
substances find their way to the lake during runoff. Nutrients are also
contributed in significant quantities from animal wastes.

If an estimated rate of 250 pounds of total phosphorus per square mile per
year is used to calculate the agricultural contribution, almost six million pounds
are contributed to Lake Erie per year from this source. The nitrogen input from
runoff is at least ten times this amount.

At least eight million tons of silt are discharged to Lake Erie from agricultural runoff each year. Nearly half of this is discharged to the western basin.

The Maumee River is the greatest contributor to rural runoff pollution, in both nutrients and silt, in the Erie basin. About two million tons per year of nutrient-laden silt enter the lake from this drainage basin.

It is not likely that the rate of silt input from rural runoff will increase by any significant amount in the future because of improved soil conservation practices. In fact, it appears that the present rate is less than it has been in the past. The inputs of nutrients from rural runoff are likely to increase however, because of the rapidly increasing usage of fertilizers.

Combined Sewer Overflows

Combined sewer systems are recognized as very significant sources of pollutants both to tributaries and to Lake Erie. The more important materials contributed are BOD, bacteria, and the nutrients, nitrogen and phosphorus. Beaches are closed in many places because of the bacterial loadings.

Many large cities in the Lake Erie basin have combined sewer systems carrying both sewage and surface drainage water. During dry weather the sewer systems supposedly direct all flow to a sewage treatment plant. During periods of precipitation the excess flow bypasses the treatment plant and goes directly to the nearest watercourse. Many of the systems are in such poor condition that sewage is continuously bypassed; Cleveland's system is a prime example.

At present the largest contributors to Lake Erie pollution from combined sewers are the cities of Detroit, Cleveland, and Toledo. These have an immediate detrimental effect, particularly at bathing beaches in the vicinity. Combined sewers are the main reason that the entire shoreline in the Cleveland metropolitan area is unsafe for swimming.

Approximately 40 billion gallons per year flow from combined sewers directly to the basin's waterways. About half of this flow is untreated municipal waste, i.e., sewage bypassed to the overflow during rainstorms. Sewer overflows, on an annual average, contribute wastes equivalent to the oxygen demand of raw sewage from approximately 600,000 persons. *The wastes from combined sewer overflows will be much greater in the future if controls are not instituted now.* Tables 6 and 7, for example, show the projected contribution of oxygen-demanding substances from storm water overflows in 2020.

Sewer separation is also favored by the fact that the phosphorus contribution from combined sewers is over four times greater than that from separate sewers. Thus, in 2020, if sewers are not separated the total phosphorus load, now approximately 6,000 lbs/day, will increase to about 20,000 lbs/day. With separation, the 2020 load would be less than the present-day load.

Other Sources of Wastes

Vessel Wastes. Vessel discharges are locally damaging, especially in harbor areas, although they are not a significant factor in the overall water quality of the lake. The bacterial and nutrient pollution load from commercial vessels on Lake

TABLE 6. Projected BOD in lbs/day from storm water overflow
assuming no further control measures or separation of existing sewers.

Subarea	1967	2020
Southeast Michigan	43,000	90,000
Maumee	11,000	23,000
North Central Ohio	4,000	8,000
Greater Cleveland-Akron	30,000	65,000
Northeast Ohio	500	2,000
Pennsylvania	2,000	8,000
New York	12,000	25,000
Total	102,500	221,000

TABLE 7. Projected BOD lbs/day from storm water overflow* assuming
separate sewer systems throughout.

Subarea	2020
Southeast Michigan	38,000
Maumee	9,000
North Central Ohio	3,000
Greater Cleveland-Akron	26,000
Northeast Ohio	2,000
Pennsylvania	2,900
New York	12,000
Total	92,900

*These tables show that the combined sewer overflow in 2020 would amount to 221,000
lbs/day BOD—more than twice the present discharge. But, if sewers are separated this
discharge will amount to only 92,999 lbs/day in 2020, less than the present discharge of
102,500 lbs/day.

Erie is equivalent to the raw sewage of 1,200 persons for eight months of the
year or a permanent population of 800. The pollution contribution from
pleasure craft is equivalent to the raw sewage of a permanent population of
5,500. Areas of particular concern from vessel wastes are around Detroit,
Toledo, Sandusky, Vermilion, Rocky River, Cleveland, Ashtabula, Erie, and
Buffalo.

Shore Erosion. Erosion of the lake shore contributes an estimated 16 million
tons of silt to Lake Erie per year and adds to the nutrient problem. Shore erosion
is responsible for most of the nearshore turbidity of the lake during storms.

Dredging Dumps. All harbors along the United States shore are dredged
periodically to maintain navigation channel depths. In most harbors the removed material is a combination of silt and municipal and industrial wastes,
amounting to some six million cubic yards (Table 8) annually in the Erie

TABLE 8. 1967 estimated harbor dredging spoil dumped in Lake Erie.

Harbor	Volume cu. yds.	Principal sludge source
Monroe (Raisin River)	250,000	Industrial
Bolles Harbor	186,000	Rural
Toledo (Maumee River)	1,000,000	Rural
Sandusky	600,000	Rural
Huron	180,000	Rural
Lorain (Black River)	500,000	Industrial, municipal, rural
Rocky River	60,000	Municipal, rural
Cleveland (Cuyahoga River)	1,300,000	Industrial, municipal, rural
Fairport (Grand River)	360,000	Industrial, rural, municipal
Ashtabula	350,000	Industrial, rural
Conneaut	400,000	Rural
Erie	200,000	Municipal
Dunkirk	26,000	Industrial
Buffalo	620,000	Industrial
Total	6,032,000	

basin, more than one-fourth of the total silt load to Lake Erie. Cleveland Harbor requires more dredging than any other harbor on the Great Lakes, and it contains some of the most noxious materials.

In many cases dredging wastes harm water quality by the addition of BOD and nutrients. It has been the policy to dump the dredged materials in the lake within a few miles of the dredging sites, which may be transferring highly polluted substances to relatively unpolluted areas.

Construction Runoff—Highway and Urban Development. Pollution from construction sites is mainly silt and is similar to agricultural runoff, but the rate per unit area is much higher. The problem is becoming increasingly serious because of the recent intensification of highway and housing programs in the Erie basin covering large areas of land. There apparently has been no adequate program of reseeding, catch-basins, etc. during construction; and the land is left barren for long periods, especially over winter.

Trash and Debris. The waterways of the Lake Erie basin are being used, in many places, for the disposal of solid wastes. Municipal and industrial dumps presently exist along the banks of most major streams. These dumps are eyesores and contribute oils, oxygen demand, trash, and other wastes.

Fallen trees and stumps choke rivers and streams at many places, blocking flow and collecting floating material. During high flows the debris is flushed downstream to harbor areas and to the lake where it interferes with commercial navigation and small boating.

Oil Discharges. Most harbors and some other areas of tributaries are subjected to occasional spilling or dumping of oil, either from vessels or shore installations. Some industries, such as oil and steel companies, discharge oil intentionally and continuously. These occurrences are not only disgraceful and a direct fire hazard, but are exceedingly harmful to wildlife, waterfowl, fish, and other water supplies, and recreation. The persistence of oil, being, for the most part, nondegradable naturally, makes this problem a difficult one to overcome. The possibility of a major cargo spillage within the lake proper is real and poses a disastrous threat to the entire lake.

Methods of coping with oil pollution after a discharge, either locally or basin-wide, are not now adequate, but efforts are being made to correct this situation, especially by the Coast Guard and FWPCA.

Potential Sources of Pollution. Offshore drilling in the lake proper for oil and gas is now being contemplated by several companies on bottom land to be leased from the states of Ohio, Pennsylvania, and New York. This is a potential source of pollution, mainly from oil spillage, brine, and drilling muds. If considerable quantities of oil are encountered in drilling, the risk of oil pollution will be of the highest degree and could lead to disaster.

Reactor plants for power sources seem to be inevitable. These represent potential sources of radioactive pollution substances. Their main contribution to the degradation of Lake Erie is and will be thermal pollution because of the necessarily great amount of cooling water used.

Very important potential sources of pollution are the ultimate disposal sites of the residue from waste treatment plants. This is especially important in regard to nutrients. Nutrients removed at a treatment plant can have little effect on improving Lake Erie water quality if the nutrients still are discharged to the drainage system.

Federal Activities. In the Lake Erie basin there are 15 Federal installations that discharge directly to public waters. Of these, five are considered not to have adequate facilities:

1. The Selfridge Air Force Base in Southeast Michigan operates a secondary waste water treatment plant which discharges a chlorinated effluent into the Clinton River. There are some base facilities served by septic tanks followed by sand filters that occasionally overflow to the river without disinfection. Plans are underway to connect the base's system to the Detroit Metro System.

2. The Naval Air Station at Groselle, Michigan has been responsible for significant quantities of oil in Frenchmen's Creek attributable to aircraft washings and dumps of engine oil. This can seriously interfere with wildlife and boating recreation. The Navy plans to move its air facility to Selfridge Air Force Base by July 1, 1969.

3. The Lewis Research Center (NASA) Plum Brook Facility, located near Sandusky has a complement of approximately 700. A secondary treatment plant with trickling filter and final coagulation for phosphorus removal has been constructed for treating all wastes. Tests are presently being conducted to establish maximum removal of all constituents, including phosphorus.

4. The Michigan Army Missile Plant at Warren, Michigan provides the equivalent of secondary treatment for all domestic and industrial wastes which have a combined volume of 0.8 mgd and an average BOD removal of 94%.

Consideration is presently being given to connecting to the Detroit Metro System to comply with additional State requirements.

5. The Detroit Arsenal in Warren, Michigan presently discharges its municipal wastes to the Warren Sewerage System. Additional treatment is required for some industrial and coal storage drainage waters for the removal of oil and control of pH.

For these five Federal installations, adequate treatment facilities (complete secondary treatment with phosphorus removal for all domestic wastes is considered the minimum treatment necessary) should be provided by 1968. It is recognized that programs are in effect by the Air Force, Navy, and NASA to correct these inadequacies.

Constituents in Waste

The waste substances that are discharged to the lake from municipal and industrial outfalls, tributaries, and land drainage are many; and their effects on water uses are varied. Many substances have severe effects on water uses in the localities of the discharges. The more important of these are acid, oil, cyanide, iron, coliform bacteria, phenol, and oxygen-consuming materials.

The most damaging substances affecting the total waters of Lake Erie are nutrients. Nutrients given the most attention are nitrogen and phosphorus, because, following carbon, they are required in the greatest amounts for the production of aquatic plants. Controlling one of these substances would control the rate of overenrichment in Lake Erie. Phosphorus is singled out in the discussion which follows because it is the one nutrient most amenable to treatment and control. *Other substances having damaging effects on the total waters of the lake are suspended solids (sediment), and carbonaceous oxygen-consuming materials.* Chlorides and dissolved solids have not reached damaging concentrations, but their dramatic increases indicate the rate at which water quality has been degraded. Summaries of the major known sources and loads of suspended and dissolved solids, chlorides, BOD, and phosphorus to Lake Erie are presented in Table 9.

Suspended Solids

Damage to Lake Erie from suspended matter is dependent on the nature of the material. Suspended matter from municipal discharges is primarily organic and oxygen-consuming, and its deposition results in enriched bottom muds or sludge banks. Effects of these wastes are largely local and can be corrected by proper treatment. Suspended matter from certain industries and the material from land erosion are largely inorganic and serve to fill harbors, embayments, ship channels, and the lake. Over 24 million tons of sediment washes into Lake Erie annually; two-thirds of this comes from shore erosion. Another 9 million tons annually is carried to the lake in dredging operations.

Principal sources of suspended solids discharged to Lake Erie are the Detroit, Maumee, Cuyahoga, and Grand Rivers which represent more than 40 million

TABLE 9. Waste loads to Lake Erie basin waters—1966 (pounds per day).

Source	BOD	Chlorides	Total phosphorus	Suspended solids[1]	Dissolved solids
Industrial	480,000	10,980,000	5,900		
Municipal	900,000	1,830,000	86,400		
Rural runoff		} 3,120,000	18,220		
Urban runoff			8,760		
Lake Huron outflow	950,000	6,500,000	<20,000	3,800,000	116,000,000
U.S. undifferentiated				73,000,000	} 84,000,000
Canada undifferentiated	100,000 est.	2,900,000	18,000	57,100,000	
Total	2,430,000	25,330,000	157,280	133,900,000	200,000,000

[1]Over two-thirds of this comes from shore erosion. Exclusive of solids (9 million tons annually) deposited in the lake in dredging operations.

pounds per day. About 1.5 million pounds per day of suspended solids of the Detroit River are from industrial and municipal sources. The Maumee discharges are largely silt from land runoff. The greatest quantities are released during periods of heavy rain and high runoff; therefore, control must be instituted through improvements of land use practices in the watershed. The Cuyahoga and Grand Rivers' (Ohio) discharges are believed to be largely from land runoff and from industrial and municipal wastes. The load in the Cleveland harbor results in severe discoloration and the need for frequent dredging.

Carbonaceous Oxygen-consuming Materials

Carbonaceous oxygen-consuming materials, usually measured by the 5-day biochemical oxygen demand (BOD), are pollutants to streams in that they depress dissolved oxygen levels. This immediate effect is not as pronounced in

TABLE 10 Present untreated municipal and industrial BOD loads by subbasin lbs/day.

Area	Untreated BOD	With recommended treatment
Southeast Michigan	1,180,000	200,000
Maumee — Indiana	90,000	6,000
Maumee — Ohio	300,000	45,000
North Central Ohio	110,000	16,000
Cleveland — Akron	470,000	70,000
Northeast Ohio	30,000	5,000
Pennsylvania	220,000	34,000
New York	140,000	24,000
Total	2,540,000	400,000

lakes such as Lake Erie because of its tremendous oxidation capacity. However, BOD is a measure of wastes that are used by bacteria in cell growth and reproduction, thereby creating sludge which settles to the lake bottom. Carbonaceous BOD of wastes is most effectively removed by secondary or tertiary treatment.

The present and projected daily BOD loading for the entire basin is shown in Figure 3 along with the loading after various degrees of reduction. Figure 4 shows projected municipal loadings for each of the subbasins. As this figure indicates, the Detroit area contributes more BOD to Lake Erie than all other known sources combined. Table 10 shows present industrial and municipal BOD

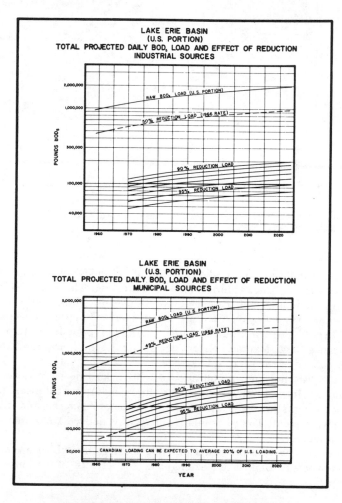

FIGURE 3. Lake Erie basin total projected daily
BOD load and effect of reduction—industrial
and municipal sources.

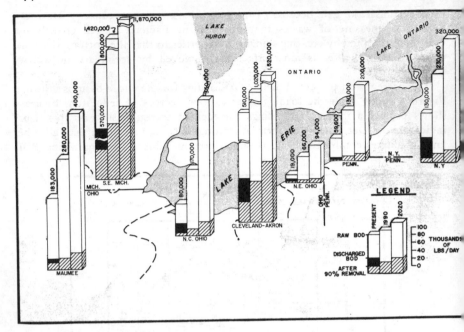

FIGURE 4. Present and projected BOD load discharged in the Lake Erie basin.

loads by subbasins of both the untreated waste and with treatment levels recommended in this report.

Chlorides

Lake Huron discharges 6.5 million pounds of chlorides per day, accounting for 26 percent of the total chloride load to Lake Erie. The Detroit-Windsor area discharges 9.5 million pounds per day or 38 percent. Thus, *nearly two-thirds of the chloride load to Lake Erie enters at the mouth of the Detroit River*. The Grand River (Ohio) contributes another 3.9 million pounds, or 15 percent; and the Cuyahoga and Maumee Rivers contribute 1.5 million pounds per day or about 6 percent of the total to Lake Erie.

Table 9 lists the known chloride loads to Lake Erie from various kinds of contributors. In addition to the Lake Huron input, contributing 26 percent, industry accounts for 43 percent, municipal wastes 7 percent, and street deicing (runoff) 12 percent. The remaining 12 percent is derived from undifferentiated Canadian sources.

Figure 5 shows the projected chloride loadings by subbasin, and Figure 6 shows projected total lake loading and the effects of various degrees of reduction.

Dissolved Solids

Dissolved solids concentrations at the head of the St. Clair River average 110 mg/1, at the head of the Detroit River 126 mg/1, and Lake Erie at Buffalo 180

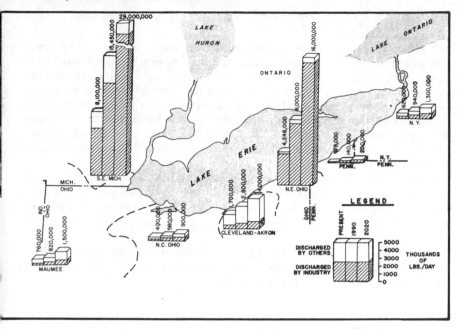

FIGURE 5. Present and projected chloride discharges in the Lake Erie basin
without additional controls.

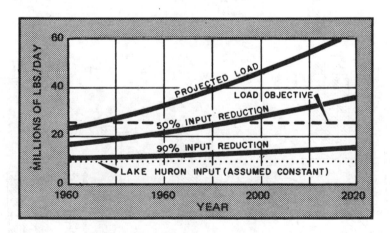

FIGURE 6. Projected chloride load to Lake Erie and
loads with various in-basin load reductions.

mg/1. These levels represent daily inputs of 116 million pounds per day from the
watershed above Detroit and a discharge of almost 200 million pounds per day
to the Niagara River from Lake Erie. *Most of the increase within Lake Erie
actually is derived from the Detroit area.*

Phosphorus

The principal in-basin sources of phosphorus are municipal wastes (72 percent), agricultural runoff (17 percent), urban runoff (7 percent), and industrial wastes (4 percent), Table 11. These inputs total 137 thousand pounds per day and exclude the Lake Huron input of somewhat less than 20 thousand pounds per day. In municipal wastes about one pound per capita per year is contributed by human excreta and 2.5 pounds per capita per year by detergents. Phosphorus from agricultural runoff amounts to about 250 pounds per square mile per year. Urban runoff contributes phosphorus at the rate of about 530 pounds per square mile per year.

Figure 7 shows the contributions of phosphorus for each of the subbasins and the projected contributions for the years 1990 and 2020. *Phosphorus loadings to Lake Erie will increase nearly 2.5 times by 2020 if the present rates continue unchecked.* Figure 8 shows total projected phosphorus inputs from various sources.

The Detroit area contributes by far the largest amount (40 percent) of phosphorus to Lake Erie, more than twice as much as either the Cleveland or the Maumee areas, the next two major sources.

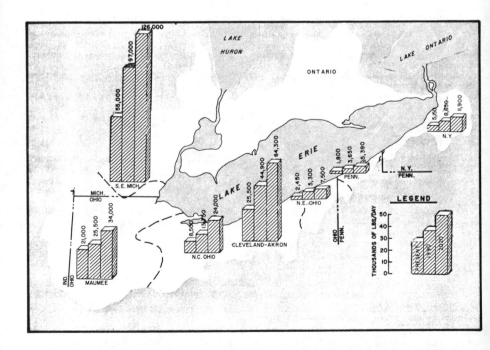

FIGURE 7. Present and projected phosphorus discharges in the Lake Erie basin without additional controls.

TABLE 11. Present and projected phosphorus discharges to Lake Erie
exclusive of Lake Huron input—lbs/day.

Subbasins	Municipal waste	Industrial waste	Urban runoff	Rural runoff	Total
Present Loading					
Southeast Michigan	46,000	3,000	3,000	3,000	55,000
Maumee River basin	9,000	1,000	1,000	10,000	21,000
North Central Ohio	3,800	500	1,600	2,600	8,500
Greater Cleveland–Akron area	22,000	800	2,000	700	25,500
Northeast Ohio	1,100	100	500	750	2,450
Pennsylvania	1,400	100	110	220	1,830
Western New York	3,000	500	650	1,000	5,150
Ontario	11,900	Unknown	450	5,500	17,850
	98,200	6,000	9,310	23,770	137,280
Projected 1990 Loading					
Southeast Michigan	85,000	4,500	4,500	3,000	97,000
Maumee River basin	12,000	1,500	2,000	10,000	25,500
North Central Ohio	8,000	750	2,400	2,600	13,750
Greater Cleveland–Akron area	40,000	1,200	3,000	700	44,900
Northeast Ohio	3,700	200	700	700	5,300
Pennsylvania	3,100	180	160	210	3,650
Western New York	6,100	750	1,000	1,000	8,850
Ontario	21,400	– –	810	6,500	28,710
	179,300	9,080	14,570	24,710	227,660

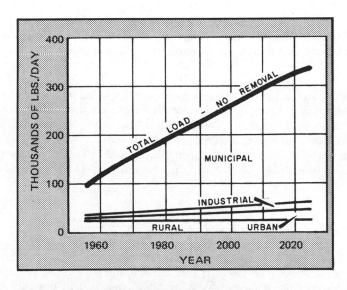

FIGURE 8. Projected phosphorus load
to Lake Erie by source—cumulative.

Radioactive Substances

The content in Lake Erie of radioactive wastes is increasing, primarily from atmospheric fallout, nuclear power plants, and from medical uses. Limits for discharge are set by the Atomic Energy Commission. The levels in Lake Erie and in tributaries are low and are not known to be harmful, but the substances are cumulative and lost only by long natural decay. Increasing conversion of the energy for electric power production from fossil to nuclear fuels indicates that radioactive waste levels will continue to rise.

Toxic Substances

Many substances toxic to aquatic life and even to human life are discharged to the waters of the Lake Erie basin. Fortunately all are either degraded or diluted to acceptable levels quickly upon reaching the lake proper.

Toxic metals such as copper, cadmium, chromium, lead, nickel, zinc, and iron are discharged in significant quantities by primary metals and metal-fabricating industries. Areas of concern are near the mouths of the Rouge, Black, Cuyahoga, and Buffalo Rivers.

Some toxic metals, such as copper, may be accumulated and stored by algae and cause difficulties in the life forms which consume these organisms.

Many organic chemicals are toxic, or near toxic, such as the multitude of insecticides, and the organic compounds discharged by the plastics industries. Phenols are waste products of the iron and oil industries. The Detroit-Maumee basin, Lorain, Cleveland-Akron, and Buffalo areas, produce large quantities of organic chemical wastes.

Acids

Acids, products of many industries, are indirectly toxic in that they may reduce the pH of the receiving waters to levels interfering with fish and aquatic life. Detroit, Cleveland, and Buffalo are areas of large acid discharges.

Oil

Oil is persistent and difficult to decompose in water and has serious effects on all forms of water uses. Large quantities are discharged continuously by the steel and oil industries in Detroit, Toledo, Cleveland, and Buffalo. Oil is also contained in municipal wastes in significant quantities.

Oil, as a heavy floating scum, is continuously present in many harbor areas, particularly in the Rouge at Detroit, the Cuyahoga at Cleveland, and in the Buffalo River.

Heat

Heat or thermal pollution is a waste product of many industries. It is significant in electric power production. The increase in temperature may be $20°F$ or more with one pass through a power plant. This would be significant in small bodies of water where the water is recycled. It can affect temperature-sensitive organisms in the lake in the immediate vicinity of the discharges, but it does not cause a measurable temperature increase in the lake as a whole. It is possible though not likely, that increased power production from nuclear energy will cause a measurable increase.

Steel industries also cause local rises in temperature with cooling water and slag discharges. This source of heat is not likely to increase significantly except locally in tributaries. It is in tributaries that temperature problems are now of concern.

Agricultural Sources of Water Pollution

Berlie L. Schmidt*

Introduction

With the great public interest in water and air pollution and environmental quality, many statements and questions have been raised concerning the role of agriculture as a source of pollutants to the environment. Many statements have been made without factual basis, particularly by persons not acquainted with agricultural research in resource management. Although much research has been started in recent years to study the possible role of agricultural pollutants, there has not been sufficient data and factual evidence to properly answer many of the questions raised concerning the role of agriculture in water pollution.

Types of Pollutants from Agricultural Cropland

There are three classes of potential pollutants from agricultural cropland. These include (1) sediment, (2) pesticides, and (3) chemical fertilizers.

Sediment may be one of the most costly and serious of all pollutants from agricultural cropland, although it is often overlooked. The importance of sediment as a pollutant comes from: the high costs resulting from sedimentation and filling ·of streams, causing reduction in carrying capacity, and increased possibility of flooding; high dredging costs of sediments carried into waterways, lakes, and shipping channels; and the fact that sediments are usually composed of the finer-textured materials from our agricultural soils, which are generally higher in organic matter and nutrient content than the average topsoil in agricultural watersheds.

Pesticides include the insecticides and herbicides used on agricultural land. These materials have come under close scrutiny and criticism as possible pollutants in recent years.

*Department of Agronomy, The Ohio State University and Ohio Agricultural Research and Development Center, Wooster, Ohio.
Edited and reprinted from a report to the Panel on Agricultural Contribution on Water Pollution, Upper Sandusky, Ohio, 7 January 1970.

Chemical fertilizers include plant food materials, particularly nitrogen, phosphorus and potassium, which may act as potential pollutants since they greatly stimulate aquatic plant growth in lakes and streams, and cause eutrophication or enrichment of our water bodies. The two nutrients of most concern to water pollution presently are nitrates and phosphates.

Sediments

Sediments not only cause filling of streams and lakes, but, as mentioned previously, are usually composed of fine clay materials high in colloidal material or organic matter, which may carry adsorbed nutrient ions such as phosphorus, potassium, and organic chemicals such as pesticides. According to the Federal Water Pollution Control Administration's Lake Erie report (FWPCA, 1968), nearly eight million tons per year of silt are deposited in Lake Erie from agricultural watersheds. About half of this goes into the Western Basin, with about two million tons per year coming from the Maumee River Basin alone. This silt comes from not only surface erosion, but also from stream bank erosion, construction sites, highway and urban areas.

Although the soils of northwestern and northern Ohio are relatively flat and low in percent slope, these soils are high in clay content and nutrient status. Also, a high percentage of these soils are intensively cultivated, and are left bare during critical erosion periods of the season, resulting in the possibility of significant amounts of fine colloidal material being removed by surface runoff and erosion. Much past data have shown that the use of recommended soil and water conservation practices on these agricultural watersheds will greatly reduce runoff and erosion and go far in reducing possible pollution from sedimentation and other materials.

Agricultural Chemicals

Agricultural chemicals, including modern pesticides and chemical fertilizers, are essential for efficient crop production. However, they can be potential pollutants if they are not used according to recommendations or with poor land use practices. These materials can be carried off of crop land in two forms: (1) soluble forms in runoff or in percolate water, (2) insoluble forms or attached to colloidal sediments. Many of these materials, particularly those in the form of positively-charged cations, are adsorbed on clay or colloidal particles and carried off with sediment. The clay minerals involved in Ohio soils include illite, montmorillonite, and kaolinite. The montmorillonite form of clay mineral is particularly active in adsorbing these chemicals, since its lattice structure can expand or shrink according to the amount of water held in the mineral. When this mineral lattice structure is opened up by the presence of water, extra surfaces are available for adsorbing and holding chemical ions. Sediments removed by surface runoff often contain a large percentage of these clay minerals as colloidal material, and are an important method by which these chemical materials can be removed from agricultural soils to become potential pollutants.

Fertilizer Nutrients

Fertilizer nutrients, especially nitrogen, phosphorus, and potassium, may also act as nutrients for aquatic organisms and help speed the eutrophication of lakes. Nitrogen and phosphorus in runoff are particularly of concern in water pollution and algae bloom.

Nitrogen in the nitrate form is water-soluble and moves with soil water through the soil profile. Nitrates have been particularly of concern because of their connection with methemoglobinemia in infants, which reduces the capability of the blood to carry oxygen. Nitrogen also increases aquatic plant growth, which, upon decomposition, decreases the oxygen level of the water and affects fish populations.

Phosphorus also causes eutrophication of lakes by stimulating aquatic plant growth and the growth of water organisms. Phosphorus, however, unlike nitrogen, does not move with water in the soils, but is attached strongly in the soil colloids. Therefore, it remains fixed and concentrated near the soil surface, which in turn can make it more susceptible to removal by surface erosion.

Higher concentrations of these major plant nutrients are generally found in the soil surface than deeper in the profile. These materials are selectively removed in eroded sediment, which is therefore higher in plant nutrients, organic matter, and colloidal clay materials than the normal soil. Enrichment ratios of plant materials in sediments, or in other words, the ratios of the content of these materials in sediment to that in normal soil, are shown by the following examples: Organic matter — 2.1, nitrogen — 2.7, available phosphorus — 3.4, and exchangeable potassium — 19.3 (Barrows and Kilmer, 1963). Therefore, the reduction of sediment removal from soils would also significantly reduce the amount of nutrients being removed from agricultural cropland into nearby waterways.

Sources of nitrogen and phosphorus into Lake Erie were recently compiled by the Federal Water Pollution Control Administration in their report on Lake Erie in 1968. According to this report, the percentages of phosphates coming into Lake Erie from different sources are: municipal — 72%, industrial — 4%, urban runoff — 7%, rural runoff — 17%. Household detergents made up about 66% of the phosphorus coming from municipal sources.

A large percentage of the nitrates coming into Lake Erie were, according to the FWPCA report, from sewage and municipal wastes, although agricultural watersheds and chemical fertilizers were mentioned as potential nitrate sources.

Agricultural sources of nitrogen and phosphorus can be not only chemical fertilizers added to the soil, but also the natural amounts of these nutrients contained in normal fertile topsoil material. For example, nitrogen contained in the surface foot of soil is somewhat higher in northwestern Ohio than in the remainder of the state, with an overall average of between 0.1 and 0.2 percent nitrogen, or about two to four thousand pounds per acre of total nitrogen. Much of this nitrogen is held in organic matter in the A-horizon or topsoil layer, and is only made available through decomposition and mineralization. A recent survey of average amounts of nutrients applied as fertilizer to soils in northwestern

Ohio (Kroetz and Schmidt, 1969) shows an average application of N, P_2O_5, and K_2O of 150, 107, and 140 pounds per acre on the farms ranking in the highest quartile of crop yields. Farms with average yields in the lowest quartile had average annual applications of 78, 80, and 74 pounds per acre of these nutrients. Much of the nitrogen applied as fertilizer is removed from the soil by uptake and removal by crops, and losses by volatilization.

Phosphorus, both that in the native soil profile and that applied as chemical fertilizer, is rapidly fixed to a large extent in unavailable forms by soil colloids and the organic matter fractions in the soil profile. Again, much of the phosphorus applied as fertilizer barely fulfills crop demands. Little is available for removal by soil water. Because phosphorus is concentrated near the soil surface in organic matter and on soil colloids, it is susceptible to removal with the sediment in surface erosion and runoff, but past data have shown very small percentages of the total phosphorus in soils being removed from agricultural croplands in runoff.

In general, past research data have shown that the nutrients applied in chemical fertilizers have been a very minor source of possible pollutants in surface runoff. As shown in table 1, the amount of soluble phosphorus removed from cropped watersheds at Coshocton, Ohio, was essentially the same as from unfertilized wooded watersheds. Nitrate nitrogen was slightly higher from these cropped watersheds than from the wooded watersheds, but again at relatively low levels. Essentially the same data was shown from cropped watersheds for phosphorus and nitrogen in Morris, Minnesota.

Studies underway on fertilizer materials coming from tiled drains on agricultural soils in northcentral Ohio have shown essentially the same levels of nutrients coming from tile effluent as found in the normal Sandusky River water as shown in table 2.

One potential source of pollution from agricultural land, however, might be

TABLE 1. Nutrients in runoff from agricultural watersheds.

		Content in runoff (lbs/A/yr)
Coshocton, Ohio:		
Wooded watershed, Sol. P	—	0.03 - 0.06
Cropped watershed, Sol. P	—	0.03 - 0.06
(40 lbs P/A/yr applied)		
Wooded watershed, NO_3-N	—	0.5
Cropped watershed, NO_3-N	—	3.0
Morris, Minn.:		
Cropped watershed,		
P	—	0.06 - 0.2
N	—	0.7 - 3.0

(Wadleigh and Butt, 1969)

TABLE 2. Chemical composition of tile effluent
and Sandusky River water, 1958-1968.

| | Content, in ppm | |
Chemical	Tile effluent	Sandusky River
Calcium	65	74
Magnesium	39	26
Sulfate	120	125
Chlorides	14	14
Nitrates	12	7

(Taylor, et al. 1969)

from livestock feedlots. Runoff from these areas can be high in organic materials, as well as nitrates and phosphates. These sites are purposely built on sloping land for good drainage, which increases the amount of runoff of surface water capable of carrying these materials into nearby waterways and streams. Soils under livestock feedlots have been found in some cases to be significantly higher in nitrates than under normal cropland, as shown in table 3.

Data from the University of Missouri have shown amounts of nitrate nitrogen as high as 500 to 600 pounds per acre foot at the five to ten-foot depth in the soil under livestock feedlots (Smith, 1967). To solve this problem of agricultural feedlot runoff, such methods as settling basins, diversion ditches, or primary treatment systems have been proposed and look promising for reducing potential pollution.

Data reported from the chemical analysis of samplings from rivers, streams and lakes have often been used as indications of pollution from agricultural land. However, definite conclusions cannot be made from such data as to the source of these materials, since large amounts of nitrogen, phosphorus and other nutrients, as well as pesticides and organic chemicals, can be adsorbed and held in the colloidal material and sediments on the bottoms of streams and waterways. These materials may have come from upstream industrial, municipal,

TABLE 3. Total NO_3-N in 20-ft. depth
under various land uses
(Colo., 1968).

Land use	Total NO_3-N in 20-ft. profile (lbs./A)
Alfalfa	79
Native grassland	90
Cult. dryland	261
Irrigated fields (no alf.)	506
Corrals	1,436

(Smith, 1967)

or other sources, and become fixed by the sediments from the water. Natural turbulence in the streams and the water can cause these materials to be redissolved or recirculated, and picked up in the samplings made of the stream. The only accurate way to determine the actual contribution of agricultural cropland of pollutant materials is to sample runoff directly on the agricultural watersheds themselves, including not only surface runoff but subsurface interflow as well.

Possible Misconceptions About Chemical Fertilizer Pollution

Many inaccurate statements and conclusions are made concerning pollution from agricultural sources because of a lack of understanding of the facts and data available. The following points are important to consider in the subject of potential agricultural pollution:

1. The amounts of nitrogen, phosphorus and potassium applied as chemical fertilizers in most cases barely balance crop use and removal of these materials.
2. Only a very small percentage of the total nitrogen, phosphorus, and potassium in the soil is in the available or exchangeable form.
3. The use of relatively high recommended rates of chemical fertilizers may often actually decrease the amounts of fertilizer nutrients removed from a watershed by runoff. . . . The application of recommended rates of chemical fertilizers often greatly increases the amount of plant growth and, therefore, the amount of vegetative cover, which protects the soil against raindrop impact, reducing runoff and erosion significantly. This in turn reduces the amount of sediment and pollutants removed from the watershed.
4. Large amounts of nitrogen are often fixed by symbiosis by blue-green algae or other aquatic plants living in natural waterways or lakes.
5. Most of the phosphorus in agricultural soils is held immobile in the soils and cannot move in water.
6. Many natural soils are high in their content of plant nutrients. Often, runoff from naturally fertile watersheds is as high or higher in plant nutrients than in well-fertilized cropland.
7. The soil is a very effective filter for plant nutrients or organic wastes applied on the soil surface. Research at Pennsylvania State University, the University of Illinois, and other institutions have shown the soil to be a very good potential means for disposing of city sewage and industrial wastes containing high amounts of nutrient materials because of the ability of the soil profile to filter out these materials from percolating water.

Control Practices to Reduce Potential Agricultural Pollution

Soil and water conservation practices are one of the most effective principal controls for reducing potential pollution from agricultural watersheds.

According to the FWPCA report on Lake Erie, development of 27 watershed projects in the Maumee River Basin under Public Law 566 could reduce the sediment coming from agricultural sources into Lake Erie by as much as 80%. Conservation practices have been proven to be very effective in reducing water runoff and erosion, and in this way could greatly reduce the removal of agricultural chemicals, sediments, and other potential pollutants with runoff. Such methods include: contouring, cover crops, use of crop residues, strip cropping, terracing, diversion ditches, reduced tillage, etc.

Another very effective means of soil and water conservation that may be of practical use in the intensively-cropped northwestern Ohio soil areas, is the use of the no-tillage method for planting row crops, resulting from current research at the Ohio Agricultural Research and Development Center. With this method, the soil surface is left undisturbed, with crop residues from the previous year remaining intact. Corn, soybeans, or other row crops are planted directly through the undisturbed mulch by the use of special no-tillage planters. The mulch left on the soil surface acts as a cushion for the soil against raindrop impact, and also greatly reduces surface runoff and erosion. A recent study on steeply sloping watersheds at Coshocton, Ohio over a three-year period showed soil losses of 5,823 pounds per acre from a watershed with conventional plow-disk-harrow tillage for corn, as compared to only 118 pounds per acre of soil loss from a similar watershed with no-tillage used (Harrold, et al, 1967). Similar studies in previous years at Coshocton have shown that the use of good conservation practices significantly reduces the amount of soil loss and erosion. For example, prevailing practices in the area, including up-and-down-hill cultivation and poor soil management, resulted in 8.6 tons per acre of soil loss per year, as compared to only 2.2 tons per acre per year with good conservation practices, including contouring and good soil management.

The use of soil management practices, such as tillage, crop residues, etc., which increase the amount of water infiltration into the soils, result in: greater water intake, greater crop growth, more canopy and vegetative residue, less splash and sheet erosion, less runoff, and therefore less potential pollution or sediment removal from soils. Cropping systems and soil conditions can have a large effect on the amount of water infiltration or runoff from soils.

Current and Planned Research on Agricultural Pollution Control

There is a great need for much more data on the amount of pollution actually occurring from agricultural sources in Ohio, and on the effectiveness of various pollution control methods. A number of research projects are currently underway or being planned to obtain such data, and to help plan future controls of possible pollution.

References

1. Barrows, H. L., and V. J. Kilmer. 1963. Plant nutrient losses from soils by water losses. Adv. in Agron. 15:303-316.
2. Harrold, L. L., G. B. Triplett, Jr., and R. E. Youker. 1967. Watershed tests of no-tillage corn. J. Soil and Water Conservation 22:98-100.

3. Kroetz, M. E., and W. H. Schmidt. 1969. 1968 Corn and soybean survey in Northwest Ohio. Ohio Cooperative Extension Service, Ohio State University. Mimeo.
4. Smith, G. E. 1967. Fertilizer nutrients as contaminants in water supplies. *In* Agriculture and the quality of our environment. Amer. Assoc. Adv. Sci. Pub. 85, p. 173-186.
5. Taylor, G. S., J. H. Wilson and A. P. Leech. 1969. Chemical analysis of tile drainage water at the Tiffin, Ohio drainage experiment. Ohio Agr. Res. and Dev. Center Agron. Mimeo. Series 208.
6. U.S. Department of the Interior, Federal Water Pollution Control Administration. 1968. Lake Erie Report—A plan for water pollution control. FWPCA, 107 p.
7. Wadleigh, C. H., and C. S. Butt. 1969. Conserving resources and maintaining a quality environment. J. Soil and Water Conservation 24:172-175.

Part Three

Geologic Controls and Ground-Water Pollution

T
he geologic framework, the rocks that make up an underground reservoir, controls to a large extent the movement of the enclosed fluids. How aquifers become contaminated and the spread, dilution, direction of movement and velocity of ground water are described in the following section by H. E. LeGrand, Morris Deutsch, and R. H. Brown. Monitoring changes in ground-water quality is discussed by H. E. LeGrand. These reports should provide sufficient background material for those interested in the mechanics of ground-water pollution and furnish a scientific basis for water planning and management, including legal implications. In addition, they are intended to serve as background material for Part Four, Examples of Ground-Water Pollution.

Only recently have public agencies begun to study and evaluate the ecological effects of solid-waste disposal. By considering the geologic and hydrologic framework of a potential disposal site, it should be possible to design landfills that will have a minimal effect on the chemical and biological quality of adjacent water resources—both surface and underground. Considerations such as these appear in the report by W.J. Schneider.

Photo opposite courtesy of U.S. Department of Agriculture.

Environmental Framework of Ground-Water Contamination

H. E. LeGrand*

Introduction

Although ground-water contamination appears to occur willy-nilly in the hydrogeologic environment, geologists and hydrologists have not in general been aggressive in occupying this aspect of the domain within the limits of their disciplines. In the past, many problems of ground-water contamination seemed to have developed casually, and to the geologist and hydrologist, busy in other work, these problems may appear isolated and easy to evaluate—at least in hindsight. Consideration of ground-water contamination as a multitude of independent problems, separately solvable as each arises, is outmoded; wise policies, relating water supply to contamination, are needed to alleviate and to forestall problems.

Contamination, as referred to in this report, includes any deterioration of the quality of ground water. It includes disposal of wastes in the ground and any other actions by man that result in the movement of undesirable water into storage space which naturally would be occupied by ground water of acceptable quality. The purpose of this paper is to place in perspective some major hydrogeologic factors for those interested in problems in ground-water contamination.

Even the apparently simple contamination problems have complex ramifications, extending into aspects of water supply, of excessive health risks, and of unwanted costs. Management and planning are key considerations, because most problems do not merely go away and because they are not solvable by simple decree. There are reciprocal changes in underground storage of contaminated and uncontaminated water, and commonly the volume of fresh ground water in many places is shrinking as wastes are dispersed by natural ground-water flow and as contaminated water spreads by pumping water from nearby wells. Ground-water contamination is not a separate entity in which its

Reprinted from *Journal of Ground Water*, Vol. 6, no. 3, 1968, pp. 14-18, with permission of the author and the National Water Well Association.

*Water Resources Division, U.S. Geological Survey. Raleigh, North Carolina.

problems can be administered or managed apart from water development problems. The interrelation of ground water and surface water is becoming increasingly appreciated, and the possibility that contaminated water may pass in either direction between a stream and adjacent ground must be recognized.

Evaluation of problems calls for appreciation of two opposing tendencies—the tendency of a contaminant to move with ground water and the tendency to be attenuated by decay or inherent decrease in potency, by chemical and physical sorption, and by dilution through dispersion of ground water. Contaminants differ greatly in their attenuation habits, resulting in a chromatographic distribution of separate contaminants where mixed wastes are involved.

Causes and Ways of Contamination

The many causes of deterioration of the quality of ground water point out the difficulties that man has of preserving ground water of good quality. The following table (California State Water Resources Board, 1955) outlines various causes of water deterioration, making a distinction between new source water, or wastes, and degradation of naturally occurring water.

TABLE 1. Causes of deterioration of ground-water quality.

Contamination and Pollution
Domestic and municipal sewage
Industrial wastes
Organic wastes
Food processing
Lumber processing
Radioactive wastes (not originally listed but placed here because of potential importance)
Mineral wastes
Metal processing industries
Mining and ore extraction industries
Oil industries
Chemical industries
Miscellaneous
Cooling water
Solid and semi-solid refuse
Degradation
Effects of development, use, and reuse of water
Irrigation return water
Surface drainage
Percolation
Interchange between aquifers due to improperly constructed, defective, or abandoned wells
Interchange between aquifers due to differentials in pressure levels resulting from excessive withdrawal
Overdraft conditions
Sea water intrusion
Salt water intrusion
Salt balance

Upward or lateral diffusion of connate brines and/or juvenile water due to over-
pumping
Contamination from the surface due to improperly constructed wells
Natural causes—inflow and/or percolation of juvenile water from highly mineralized springs
and streams
Other causes—accelerated erosion
Mineralization resulting from plant transpiration and/or evaporation

---

The degradational causes in Table 1 are also considered as contamination in
the present report. Ground water may be contaminated in several different ways;
it may result from deliberate waste disposal practices, from accidental or
unintentional causes, or indirectly as a result of man's benefit from use of water
or of the ground.

Hydrogeologic Framework

Our knowledge of the hydrogeologic environment needs to be examined. The
cases in which hydrogeologic data have been inadequate for proper planning of
water and waste projects are too numerous and too obvious to describe. On the
other hand, the myriad of geologic and hydrologic facts, of factors, and of
principles in many cases confuses even the experienced hydrologist for he is
faced with the difficult task of culling out extraneous details and of properly
weighing the pertinent features. In both instances, we have problems of
judgments, but we can alleviate these problems by placing the hydrogeologic
framework in the right perspective and by making useful generalizations. Among
the persons acting within the scope of ground-water contamination, there is a
wide range of knowledge, the proportion of hydrogeologists being low. Keen
hydrogeologic skills, properly integrated with chemical and engineering skills,
will be required in the future as aspects of contamination become more common
and more troublesome. In this report only a brief mention can be made of some
pertinent aspects of the hydrogeologic framework in which contamination
occurs.

Porosity and Permeability

Some knowledge of the porosity and permeability of earth materials is essential
in evaluating water and waste problems, whether the problem is of a general
nature or of a particular situation. In earth materials there is a direct relation
between porosity and transient or quasi-permanent storage of water, and there
tends to be a direct relation between permeability and the movement of ground
water. For a first-round porosity appraisal, a distinction should be made between
the two contrasting types of media—porous granular materials and dense rocks
with linear openings. Not only does the interstitial occurrence and movement of
water and liquid contaminants in granular materials differ from the channeled
occurrence and movement of water in dense rocks, but the degradation of some
contaminants is much more effective in loose granular materials than in

consolidated rocks. A complex situation occurs where wastes are released in loose granular materials that are underlain at shallow depths by rocks with linear openings.

Some contamination problems require efforts toward precise determinations of the hydrologic coefficients of permeability and of storage, but for many other cases it is adequate to make only the distinctions between good permeability, relatively poor permeability, and impermeability or extremely low permeability. We must appreciate the fact that the permeability of some clays may be many hundreds of times less than that of some sands. Other pertinent generalizations are:

1. In the vertical field, the contrast between the loose granular soil and underlying jointed rock is apparent, even to the uninitiated.
2. Zones of greater permeability tend to parallel, or coincide with, formational boundaries.
3. Where interbedded sedimentary rocks are flat or gently inclined, the changes of permeability in the vertical field are sharp and great.
4. Changes of permeability in the horizontal field, although common, are in many cases more gradual than in the vertical field.

Many of the heterogeneities involving permeability are amenable to useful generalizations but others are not; the point to be made is that water and included waste will tend to take preferred paths, flowing readily through permeable zones and shunning, or flowing with difficulty through, relatively impermeable materials.

Natural Movement and Discharge of Ground Water

Basic understanding of the natural paths and rates of ground-water movement is the starting point for evaluating problems of ground-water contamination. In many cases of contamination, the movement has been altered by man's activities, such as pumping of wells or adding liquids to the ground; knowledge or inferences about earlier conditions may guide decisions about remedial action on some contamination problems and may help responsibility for injustices in some cases.

Interdependent considerations controlling natural movement of ground water are geologic framework, climate, and topography. Brief, useful generalizations about the geologic framework are difficult to make, but we should be constantly reminded that ground water flows under the influence of gravity, following the most direct route from points of higher potential to points of lower potential, and producing the steepest pressure gradient and greatest rate of flow; to the extent that it can take preferred paths, water tends to flow through permeable materials and around relatively impermeable materials (McGuinness, 1963, p. 28). The frequency of precipitation in humid regions is sufficient to keep the water table relatively close to the ground surface, and the consequent mounding of water beneath interstream areas causes a continuous subsurface flow of water to nearby perennial streams. Thus, even the uninitiated in hydrology can get a general idea of the gross direction of movement of ground water in humid regions; in arid regions, however, the areas of natural ground-water discharge are more widely scattered, and the general movement of water may be less discernible. In arid regions some stretches of most streams are influent—that is, water and waste from the stream may seep into the ground; this is not the case with the generally effluent type of streams in humid regions under natural conditions, although, of course, pumping of wells near a stream may reverse the natural flow.

That ground-water discharge occurs in perennial stream valleys and recharge occurs in interstream areas is a well-known generalization, but the relationship between topography and natural movement of ground water is not mutually exclusive; for example, frequent and intense precipitation tends to accentuate the topographic relation to ground-water movement, whereas infrequent precipitation tends to subdue the relation; moreover, aspects of the geologic framework, such as the distribution of permeable and relatively impermeable beds, in relation to recharge and discharge areas will determine whether local or regional topographic conditions control the natural movement of ground water.

A general appreciation of natural conditions in the vertical dimension is prerequisite to understanding ground-water contamination. A major line of distinction is the water table, separating the zone of aeration from the underlying zone of saturation. There is a tendency for waste materials to be stationary in the zone of aeration except when leached or carried downward by infiltrating water from precipitation or from waste seepage. Dependence is often placed on the zone of aeration for natural pollution control because (1) air destroys or attenuates some contaminants, (2) soil materials, especially clays, have a generally good sorptive capacity for wastes, and (3) dispersion in the zone of aeration is slight in comparison to that in the underlying zone of saturation.

In the zone of saturation, water does not move uniformly because of geologic controls, and in many cases ground water at shallow depths moves faster than at great depths. At an early stage in evaluating contamination problems, it is appropriate to determine whether there is a single aquifer system or whether the water-table aquifer is underlain by one or more artesian units. Where the aquifer system is thick and complex, the following questions may require tentative answers. What is the general direction and relative rate of movement of the deeper water? Do relatively impermeable beds occur to clearly separate the aquifers? Does mineralized water occur in the deeper systems and is it confined to certain beds of formations?

Man's Influence

Man's influence on the natural ground-water regimen has several adverse effects as far as contamination is concerned.

Naturally occurring salty or mineralized water may be redistributed and may flow into fresh-water aquifers. Pumping wells near the sea or near coastal bodies of salty water may cause an inflow of salty water if the hydraulic head in the fresh-water aquifer is depressed sufficiently to cause the salty water to move toward the center of pumping. Along the coast and even inland where salty water underlies fresh-water aquifers, pumping and consequent reduction in hydrostatic pressure in the fresh-water aquifer may cause the underlying salty water to move upward. Salty water resulting from oil recovery may be stored or discharged on the land surface and contaminate to varying degrees the water-table aquifer. Even sincere attempts at safely injecting the brines back into deep salty-water aquifers may not prevent contamination through leaky well casings. Uncased oil tests and deteriorated well casings may result in some salt water moving into fresh-water aquifers. Irrigation in arid regions can result in a redistribution of salts and mineralized water in the ground; evaporation of slightly mineralized irrigation water tends to concentrate the salts in the soil, and the salts are later dissolved by irrigation water and carried into the water-table aquifer in an objectionable concentration.

Even where the natural ground-water flow system is sufficiently understood, consideration must be given to modification by man resulting from pumping of wells and from disposal of liquid wastes. Liquid disposal of wastes causes a mounding of the piezometric surface, resulting in some degree of outward radial flow and of some change in direction and rate of flow of the water. The cone of depression in the piezometric surface caused by pumping, on the other hand, may cause some contaminated water to converge toward water supplies, the chances and degree of contamination depending on positions of the wells and wastes sites in relation to the natural and imposed water-flow network. A change in the piezometric surface by man commonly has adverse effects, but under some conditions withdrawing or injecting water may be temporarily practiced in order to divert contaminated water and to prevent it from reaching a strategic spot.

Behavior of Contaminants in the Ground

Various contaminants differ greatly in their subsurface behavior. Many of them decay or are sorbed on earth materials and almost all contaminants lose some of their potency by dilution in ground water. Thus the tendency for contaminants to be dispersed or to become diluted is apparent. Because concentrations of some chemical contaminants tend to decrease only by dilution—for example, chlorides, nitrates, and certain minor elements—prediction of concentrations at specific places and at specific times is difficult.

Contamination from natural mineralized water is widespread (1) because of the widespread occurrence of contiguous fresh and salty water, (2) because the mineral matter does not merely go away, (3) because the salts are not sufficiently immobilized by sorptive processes, and (4) because dilution is not likely to be completely effective if encroachment of salty water into fresh water is progressive. Usually we consider dispersion and dilution as favorable tendencies, but this is not true where dilution is inadequate to lower the concentration of the contaminant to acceptable limits.

The potency of many contaminants decreases as time passes. The potency of others decreases by their aeration above the water table. Concentrations of radionuclides decrease distinctively with time, depending on the radionuclides involved and their respective half-lives. The survival time of some viruses is probably short, but the persistence of viruses in the ground is not well understood. Many bacterial wastes are readily destroyed in the zone of aeration.

The degree to which contaminants are retained on earth materials depends on the character of the contaminant and of the earth materials. Many of the more objectionable radioisotopes are cations that tend to have an exchange relation with cations of clay minerals (Thomas, 1958, p. 42), and much thought has been given to the reliance to be placed on the retention power of soils and other clays and sands to radioactive wastes. Clays generally have a greater sorptive capacity than sands, resulting in an inverse relation between the factors of sorption and permeability. Thus, contaminated water does not preferentially flow through most sorptive materials. In spite of increasing attention being given to studies of sorptive features of waste products, the increase in variety of waste products keeps ahead of research and synthesis of knowledge in this field.

One of the most difficult assignments in hydrology is trying to predict within narrow limits the movements of contaminants whose concentrations are decreased by dilution and by other means. We have no good means of collectively evaluating all aspects of attenuation.

Commentary

The contamination potential of a site or area may be framed in questions of (1) the extent of underground movement of the contaminant, (2) the possible hazard to a well site or to a stream, or (3) the degree of contamination at certain points from the source of contamination; the answers to any of these questions must be derived from a formulation of separate questions or problems centering on such aspects of the ground as (a) distribution of permeability, (b) distribution of sorptive materials (if it is important), (c) rate of water movement, and (d) paths of water movement. Equal knowledge of each of these aspects of the ground does not come with equal facility, and much manpower and money may be spent on one aspect without obtaining the desired accuracy. In many cases we develop false security by pursuing with great effort one aspect until we get a precise value for it, perhaps not realizing that the precision is wiped out when we must integrate the value with crudely estimated values of other aspects. The philosophy to be adopted is important because it controls the approach to solving or managing the problem, the effectiveness of the effort directed to the problem, and the adequacy of the solution or course of action. In a qualitative sense, we must avoid smugness on one hand and a feeling of futility on the other, and we must strive modestly for the best answers within the limits of time and money available.

Persons who are concerned with ground-water contamination and related water-supply problems must face the dilemma between safeguarding health on one hand and preventing inconveniences and excessive costs to develop on the other hand. We have no expressed methodology to tell us the limits to which we can extend waste-disposal and water-development practices without getting into trouble. Unnecessary restraints may cause inconvenience and excessive costs to develop, whereas liberal limits might not afford protection to health. As stated by Stone (1961, p. 163), "It is the middle ground, somewhere between liberal and strict limits, that will allow community growth and fair industrial competition without creating the threat of pollution, a threat difficult to determine." In the interests of economy of money, of space, and of time, it would be desirable to know, for example, how close we can locate each well from a waste-disposal site without fear of contamination. Without precise and simple methods to give this information we must rely on certain arbitrary standards that attempt to weigh the probability of contamination with the seriousness or consequences of such contamination.

There are so many interrelated factors and contingencies controlling the movement of contaminants in the ground that precise predictions are rarely attempted or are rarely possible. In many cases the tolerance for acceptable accuracy lies within broad limits, or only the upper or lower limit is critical. Bearing in mind that precise predictions of the dispersion of contaminants underground may be costly or foolhardy, we need to know what should be acceptable to prevent health risks.

There is a subtle tug of war between those whose primary consideration is reduction in living and development costs and those whose primary consideration is maintaining good health standards. Since much contaminated ground water is either not moving toward present well supplies or has not yet reached them, there has developed in some cases such philosophies as "out-of-sight out-of-mind" and "no one has complained yet." These philosophies have often led to grave consequences. We must not wait for complaints and for proof of contamination before taking action because:

1. Costs and inconvenience intensify when alternative water supplies and waste disposal sites are developed.
2. Health risks may not be easily avoided if we wait for actual contamination to occur.
3. Much of the contaminated water moving into fresh-water aquifers is essentially irreversible in direction of movement, resulting in a continual shrinking of the volume of fresh water available.

Geologists and hydrologists have been slow in facing broad responsibilities concerning ground-water contamination. The task is not merely one of trying to solve contamination problems after they arise; seldom are simple and finite solutions reached. Rather, the major responsibility lies in foreseeing and forecasting contamination problems so that they may be forestalled. We face a dilemma caused by tendencies of water laws to restrict or to make inflexible some individual actions and by tendencies of the complex hydrologic conditions to delicately change so that rigid laws are not realistic. Piper and Thomas (1958, p. 8) point out that "the realities of applied hydrology probably will tend toward compromise among individual interests in water or in use of water, over wider and wider areas, but the evolution of water law seems more likely to restrict than to widen the scope within which compromise will be possible." As pertinent laws are drafted and as potential or real conflicts arise between water use and contamination, geologists and hydrologists must be prepared to see that compromises and provisional decisions are proper. This calls for keen updated judgments and continual surveillance of the hydrogeologic environment. Progress will depend on interest within the geologic profession and on the development of experienced persons to study both specific problems and broad ramifications of contamination features.

References

1. California State Water Resources Board. 1955. Water utilization and requirements of California. State Water Resources Board, Sacramento, Calif., Bull. 2, v. 2, 358 p.
2. McGuinness, C. L. 1963. The role of ground water in the national water situation. U.S. Department of the Interior, Geol. Survey Water-Supply Paper 1800.
3. Piper, A. M., and H. E. Thomas. 1958. Hydrology and water law: what is their future common ground? In Water Resources and the Law. Michigan Univ. Law School, p. 7-24.
4. Stone, R. V. 1961. The way we do it; symposium, ground-water contamination. U.S. Dept. Health, Education, and Welfare, Public Health Service. Proc., p. 159-163.
5. Thomas, Henry. 1958. Some fundamental problems in the fixation of radioisotopes in solids; peaceful uses of atomic energy. In Second United Nations International Conf. on Peaceful Uses of Atomic Energy, Vienna, p. 37-42.

Natural Controls Involved in Shallow Aquifer Contamination

Morris Deutsch*

Introduction

Contamination of our shallow ground-water aquifers has become one of the more serious problems to be faced by those concerned with developing and managing the nation's water resources. The problems are very serious for a number of reasons:

(1) Shallow aquifers, which are the most important sources of water in many areas, are the most susceptible to contamination. They are susceptible to contamination by substances introduced from the surface by man. They are also subjected to contamination from natural sources because of changes in hydrologic regimen due to such activities as pumping, drilling, excavating, or dredging.

(2) Once an aquifer has been contaminated it probably will be exceedingly difficult—and sometimes economically unfeasible—to reclaim it.

(3) The public does not generally understand the principles involved in ground-water contamination, and therefore cannot anticipate the results of some common waste-disposal practices.

The chief purpose of this paper, therefore, is to present a general picture of the very broad scope of problems inherent in ground-water contamination incidents. A general understanding of the many factors involved is a requisite to effective abatement and reclamation programs.

Factors Involved in Aquifer Contamination

The major factors involved in aquifer contamination cases are:

(1) *The nature of the contaminant;* its chemical, physical, and biological characteristics—especially its stability under varying conditions. The number and

Reprinted from *Journal of Ground Water*, Vol. 3, no. 3, 1965, pp. 37-40, with permission of the author and the National Water Well Association.
*Water Resources Division, U.S. Geological Survey, Washington, D.C.

variety of potential contaminants that can enter our underground sources of supply are, for all practical purposes, limitless.

(2) *The hydraulics of the flow system* through which the contaminated liquid moves enroute to, and in, the aquifer. The contaminant may enter the aquifer directly by injection through wells, flow through open channels, by percolation through the zone of aeration, by infiltration or migration in the zone of saturation, by vertical interaquifer leakage through aquicludes, or free flow through open holes.

(3) *The natural processes* that may remove or degrade the contaminants from water while it is moving through the underground flow system until it is discharged to a well or stream. These include filtration, sorption, ion exchange, and, of course, dilution and dispersion. In the zone of aeration, oxidation and various biochemical processes are important degradation phenomena.

(4) *The physical and chemical characteristics of the geologic media* through which the liquid wastes flow. The best aquifers, in quantitative terms, are those composed of, and recharged through, the most permeable materials. Such aquifers, however, are the most susceptible to contamination. This paradox is illustrated by the relationship between the aquifer and a septic tank and drain field discharging common household effluents. If the intervening materials are very permeable as in sandy terranes, the septic tank "works" very well, but the aquifer may be readily contaminated by substances in the effluent such as detergents or other stable chemicals. If the intervening materials are not permeable, the aquifer is protected, but the septic tank will not operate efficiently. The structure of the aquifer is equally important because—for example—fluids percolating through primary interstices are subjected to more intensive degradation or removal conditions than liquids flowing relatively freely through secondary openings in dense rocks. The mineral content is a factor also to be considered inasmuch as some minerals, such as the various clays, take up some contaminants—especially the metallic ions—by exchange, whereas other minerals have no effect on the contaminants with which they come into contact.

It can be readily seen that the number of variations that can occur in aquifer-contamination cases are virtually limitless, but in order to analyze any single instance, it is necessary to at least consider all of the major factors involved.

Typical Hydraulic-Environmental Relationships

The following illustrations depict hydraulic situations involved in cases of aquifer contamination, and demonstrate the differences in removal or degradation that may be expected under certain conditions.

Figure 1 shows liquid contaminants injected into water-table and artesian aquifers. It is apparent that there is no possibility for natural processes to remove or degrade the contamination in the water before they enter or recharge the aquifer. The wastes may be diluted but the only natural treatment or degradation is by those processes that function within the already-contaminated

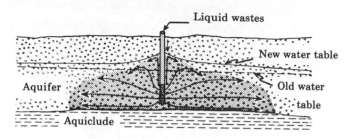

A. Water-table condition

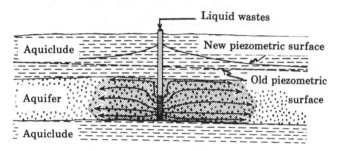

B. Artesian condition

FIGURE 1. Direct injection. There is no natural treatment before
the wastes enter the aquifer.

aquifer such as filtration, dispersion, sorption, and ion-exchange. In situations
such as these, which have actually occurred, fresh-water aquifers have been
relegated to use as waste-disposal media. This practice, of course, may be
justifiable if the aquifer is not a source of fresh water, and where the waste
cannot be adequately treated and handled at the surface such as some
radioactive or highly dangerous biological wastes. In most cases, however, such
wastes should be disposed to very deep formations, of the order of thousands of
feet.

Figure 2 shows that when a limestone or dolomite aquifer—or for that
matter any aquifer whose permeability is determined primarily by secondary
openings—is contaminated by wastes from the surface, there is again little or no
natural treatment. For all practical purposes such contamination may be
considered as direct injection into the aquifer. Many limestone and dolomite
aquifers throughout the nation have been contaminated in this manner. Where
such aquifers are the uppermost source of ground water in any area, it
commonly is difficult or impossible to determine adequate lateral or vertical
isolation distances between wells and septic tanks or other sources of contamination. In some cases contamination has spread for miles from the points of

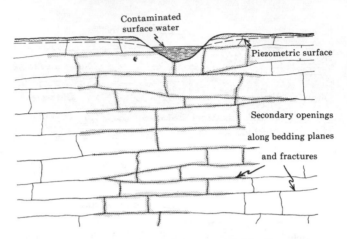

FIGURE 2. Direct injection into limestone and dolomite aquifers.
Septic tank effluents can enter such aquifers in the same manner.

injection. Dilution becomes an effectual tool to diminish the contaminant
concentration in these liquid wastes, but it is not effective in protecting users of
such waters from some types of contamination. Viruses have been of particular
concern in cases of this type in recent years.

Figure 3 shows the introduction of contaminants into an aquifer after
percolation through the zone of aeration. This is probably the safest method of
disposing of liquid wastes to the ground because the waste may be subjected to
many natural removal or degradation processes enroute to the aquifer. When
percolation is through primary interstices many common biological wastes are
naturally removed by those processes that operate under conditions of aeration.
Almost all of the suspended solid material is filtered out by the upper few inches
of soil. Within the aquifer, the contaminants are subject to natural treatment by
processes described above. However, stable contaminants in solution such as
phenolic compounds or synthetic detergents may still percolate through the
aquifer to a well. In such cases the nature of the contaminant and its reactions
with the particular natural environment involved must be considered as a
requisite to safe disposal practices.

In Figure 4 the principles involved are the same as the above, except that
resultant contamination is rarely anticipated at the time soluble solids are stored
or spread over the land surface. The leached materials in these cases usually are
stable compounds that are not readily degraded by natural processes, such as
salts or hexavalent chromium as indicated on the illustration. Common sources
of contaminants of these types are stockpiles of calcium or sodium chloride used
to melt ice and snow on public roads or discharged wastes containing chromium
compounds.

Figure 5 represents a major problem facing those water users who rely on
induced infiltration of surface water from natural sources or from artificial

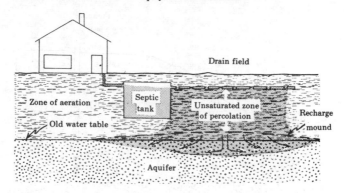

FIGURE 3. Percolation through zone of aeration. Most of the natural removal or degradation processes function under these conditions.

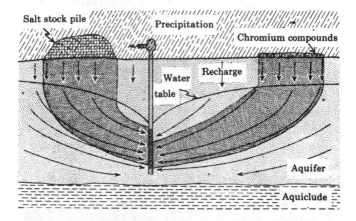

FIGURE 4. Leaching of solids at the land surface. The possibility of ground-water contamination under these conditions is rarely anticipated.

recharge facilities to replenish their aquifers. If a stream that is the source of water for a well field is contaminated by materials that are not naturally degraded under saturated underground flow conditions, "ground water" in the vicinity of the well will in turn be polluted. In all probability, however, the water after it reaches the well will be of a higher quality than the raw water in the stream because it has percolated through relatively great thicknesses of earth materials. Furthermore, some contaminants in a stream, such as cyanide, which may be lethal if ingested by man or animal, have not been found in hydraulically-connected aquifers. Because of the ever-growing use of recharge facilities in the United States, a great deal must be learned about the reactions of many contaminants contained in surface water with the earth materials in streamside aquifers.

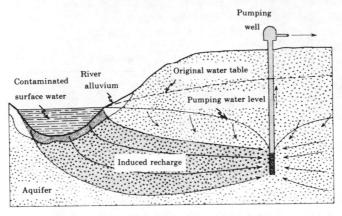

FIGURE 5. Induced infiltration from stream. Stream pollution
can result in "ground-water" pollution. Natural purification
is limited to those processes requiring no oxygen.

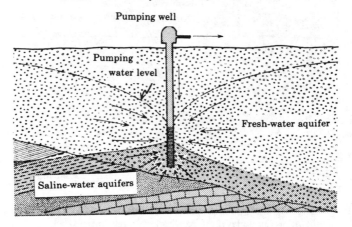

FIGURE 6. Migration from hydraulically-connected aquifer. Overpumping may
cause contaminants from natural sources to enter a fresh-water aquifer.

Figure 6 shows a possible effect of overpumping. In this case "over-pumping" indicates pumping in such an amount or manner that will result in the migration of objectionable natural waters into a shallow fresh-water aquifer. In this case the only natural degradation that occurs results usually from dilution or dispersion of the objectionable natural waters. A continuation of the pumping generally will result in a persistent increase in the degree of contamination. Migration of poor-quality water from deep aquifers might continue for considerable periods of time before it is detected. Protection against contamination of this type requires background information concerning the geochemistry ·of the aquifer tapped and those hydraulically connected with it prior to installation of well and pumping facilities. After installation, periodic monitoring of key parameters, commonly the sulfate or chloride anion, is needed to detect effects of pumping.

Figure 7 also shows possible effects of improper drilling practices. Some important aquifers in various parts of the country have been ruined by the upward flow of water from mineral-water aquifers through open holes. In most cases the holes were drilled in quest of oil, coal, or other economic mineral deposits. The existence of open holes, however, is not a requisite for contamination of a fresh-water aquifer from below. Pumping of an overlying aquifer can induce leakage through intervening aquicludes if an adequate gradient is created. Flow through open holes is more apt to cause serious contamination problems, however, because head differences required to create interaquifer movement are normally very small in comparison to the heads needed to drive water through aquicludes of low permeability.

By the same token, contamination of a shallow aquifer may eventually result in the contamination of a deeper aquifer in the manner illustrated by Figure 8. It

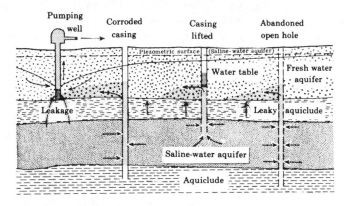

FIGURE 7. Upward leakage and flow through open holes. Some important aquifers have been ruined by improper drilling practices.

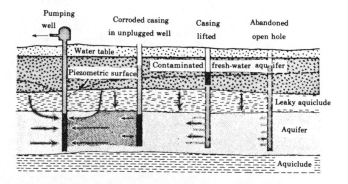

FIGURE 8. Downward leakage. Contamination of one aquifer can affect others in a multi-aquifer system.

can be seen, therefore, that if an aquifer is contaminated by any substance, development of an underlying aquifer could not be safely undertaken unless it were known that no open holes connect the two aquifers, the aquiclude would not permit significant leakage under anticipated head conditions, or that the contaminant would be removed by natural processes during flow toward and within the lower aquifer.

Conclusion

It is apparent that the problems of ground-water contamination are highly complex. All factors involved in an individual case must be considered; the hydraulics of the flow system, the chemical, physical, and biological nature of the contaminant, the natural removal or degradation processes that can be expected to operate in the underground environment, in addition to the geologic factors involved. In actual shallow-aquifer development and management programs, failure to consider each of these factors may result in the contamination of primary or even alternate sources of supply.

The technical aspects of ground-water contamination are so highly complex and involve so many scientific disciplines that solution of abatement and reclamation problems will require close cooperation and teamwork of individuals and organizations possessing competence in all disciplines involved. Unilateral approaches based on geology, chemistry, bacteriology, sanitary engineering, hydraulic engineering, and other disciplines cannot be expected to adequately cope with the problems inherent in the contamination of shallow aquifers.

Hydrologic Factors Pertinent to Ground-Water Contamination

R. H. Brown*

In 1960 Graham Walton (5) presented data concerning contamination, by sewage or other man-made wastes, of surface and underground waters. The circumstances attending the reported incidents of contamination, especially those involving ground-water supplies, have aided materially in the choice of a few principles and ideas that will identify the role of some significant hydrologic factors in the underground movement of fluid wastes.

Walton's discussion of ground water contamination refers often to physical settings into which fluid wastes are discharged at or near the land surface into cesspools, tile-drain fields, and holding ponds. Furthermore, most reported instances of ground water contamination have taken place in relatively humid environments east of the Mississippi River, where the depth to ground water is not great, frequently less than 50 feet, and where unconsolidated materials, such as sands, gravels, and clays, are the principal porous media through which the fluids must move. Thus it is convenient to imagine, for this discussion, a field environment somewhat like that idealized in Figure 1, which is a hypothetical cross section through one side of a river valley.

In Figure 1 are: the perennial stream, receiving surface drainage or runoff from the valley slopes; the water table, ranging in depth below the land surface from zero near the stream to about 50 feet and representing the positions at which water would stand in wells; the unsaturated zone, above the water table, where the pore spaces are only partly filled with water; the saturated zone, below the water table, where the pore spaces are filled with water properly called ground water; and a waste-disposal site, near the land surface. Ground water movement, as suggested by the arrows and the slope of the water table, is toward the river, and thus some natural discharge from the ground water reservoir supports the flow of the stream. The velocity of the ground water movement depends upon the properties of the porous material and the slope of the water table. If a uniform medium sand is assumed, with a water table slope

Reprinted from *Public Health Service Technical Report W61-5*, 1961, pp. 7-20.
*Water Resources Division, U.S. Geological Survey, Arlington, Virginia.

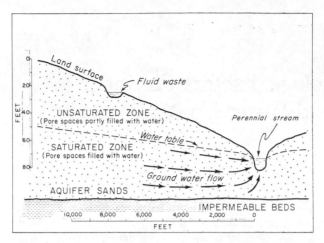

FIGURE 1. Idealized cross section of a stream valley,
showing ground water reservoir.

of 10 feet per mile, the average rate of water movement might be of the order of
a few inches per day. If the slope were doubled or tripled the rate also would be
doubled or tripled.

Fluid Movement in the Unsaturated Zone

As may be inferred from Figure 1, the unsaturated zone is an important gateway
to the saturated zone, through which precipitation that has infiltrated below the
land surface may pass to replenish or recharge the ground water reservoir. If, as
it often seems, the unsaturated zone is also the disposal site for various
contaminants and wastes, then some knowledge of how fluid moves in this zone
is timely.

Figure 2 shows the physical setting at a disposal pond or pit of shallow depth
constructed in the unsaturated zone. The porous material for both the saturated
and unsaturated zones is assumed to be uniform medium sand. The pit has been
in regular operation for many years. As fluid wastes are discharged into the pit,
some fluid seeps through the pit bottom and moves toward the water table in
the manner suggested by the vertical arrows. In this idealized setting little lateral
spreading of the fluid occurs as it moves downward; thus, most of the fluid
reaches the water table directly under the pit.

The manner of disposal into the pit greatly affects the time required for the
fluid to travel the vertical distance to the water table. Sample computations for
two hypothetical types of disposal techniques give some evidence of how much
this travel time may vary. If the disposal of contaminant to the seepage pit is
assumed to be continuous and steady, in a manner such that the rate of seepage
or infiltration is 1/2 inch per day, and no clogging occurs, fluid would never
pond in the pit inasmuch as medium sand would accept and transmit the fluid as

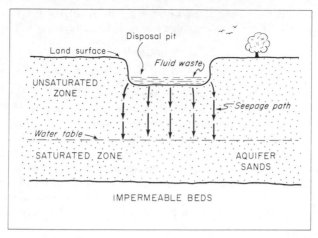

FIGURE 2. Cross section of disposal pit, showing seepage
through an idealized unsaturated zone.

fast as it arrived. Furthermore, for steady state flow the fluid would move
through virtually the entire unsaturated zone at nearly the same rate as the rate
of infiltration, 1/2 inch per day. This hypothesis may be given more rigorous
expression if it is assumed that when the sand is fully saturated, its permeability,
i.e., the rate at which water will move through it under unit hydraulic gradient
(1 foot loss in head per foot of flow path) is 70 feet per day. An equation (No.
31, p. 225) discussed by Wyllie and Gardner (6) affords a means for relating the
rate of fluid movement, in an unsaturated granular material, to the fraction of
pore space filled with the fluid. The cited equation is expressed in the form

$$K_r = \left[\frac{\theta_\varrho - \theta_o}{1 - \theta_o} \right]^4 \qquad (1)$$

where

θ_o = irreducible fraction of pore space that will retain liquid against force
of gravity

θ_ϱ = fraction of pore space filled with liquid for any given regime of
unsaturated flow

K_r = relative permeability, i.e., the ratio of the permeability for the given
regime of unsaturated flow to the permeability for saturated
flow.

In this example the rate of fluid movement in the unsaturated zone has been
assumed to be 1/2 inch per day, or 0.042 foot per day. Thus

$$K_r = \frac{0.042}{70} = 0.0006$$

In equation 1, if θ_0 is neglected because of its small value for the assumed medium sand and if the preceding value for K_r is substituted, it follows that

$$\theta_{\varrho} = (0.0006)^{\frac{1}{4}}$$
$$= 0.16$$

This means that the liquid content in the unsaturated zone need only be 16 percent of the pore space to permit fluid movement at a rate of 1/2 inch per day. Experiments with columns of unsaturated material show that, for the slow steady rate of fluid flow assumed here, this percentage can be expected to change very little throughout most of the unsaturated zone. In equation (1), if higher steady rates of waste disposal and seepage through the pit bottom are assumed (i.e., K_r is increased), the corresponding values of θ_{ϱ} will increase, but because θ_{ϱ} varies as the fourth root of K_r, it will not increase rapidly. The following tabulation gives results computed for several higher seepage rates:

Seepage, in./day	K_r	θ_{ϱ}
1	0.0012	0.19
3	0.0036	0.24
6	0.0071	0.29

Instead of the continuous waste disposal assumed for the preceding examples, it might be assumed that the same total amounts of waste are discharged intermittently. This might mean, for example, that in lieu of a continuous and steady discharge rate equivalent to a ponded depth of 4 inches per day there would be a single dumping, every other day, of fluid waste to a ponded depth of 1-1/2 feet in the seepage pit. With the data (p. 228) displayed by Baver (1), used as a guide for fluid movement through an unsaturated soil column with ponding at the surface, it seems reasonable to estimate a value of 0.7 for θ_{ϱ}. If this is substituted in equation (1) it follows that

$$K_r = (0.7)^4 = 0.24$$

From the definition of K_r, it then follows that the rate of fluid movement through at least part of the unsaturated zone is the product of (0.24) and (70) or about 17 feet per day. This rate for intermittent disposal is to be compared with rates of a few inches or perhaps a few feet per day for the equivalent continuous disposal.

The preceding computations are neither rigorous nor complete; to make them so is beyond the scope of this paper. They may be regarded, however, as approximations pertinent to significant parts of the flow system in the unsaturated zone, indicating the importance of choice of disposal technique in predicting the time required for the fluid waste to traverse the distance to the water table.

Tacit in the discussion thus far are highly idealized physical settings involving unimpeded seepage through the bottom of a disposal pit and a uniform clean medium sand. In practice these idealizations may be greatly modified by factors tending to render the disposal operations more safe. These factors may include reduction in permeability of the pit bottom by deposition of sediment in or precipitation from the fluid waste and by removal of contaminants by processes of absorption and adsorption as the fluid moves through the porous material. Evaporation and withdrawal of moisture by vegetation from the upper 8 or 10 feet of the unsaturated zone tends to lessen the amount of waste-carrying fluid that traverses the distance to the water table. In the process, however, some wastes or contaminants may be concentrated in this near-land-surface region to be picked up later and carried deeper into the unsaturated zone when natural or artificially induced infiltration of fluids increases and when adsorption sites on the previous materials at shallow depths are all occupied.

Not all disposals will occupy sites, as assumed here, with uniform medium sand having a permeability with respect to saturated flow of 70 feet per day. Corresponding permeability values for site materials varying from very fine sand or silt to coarse gravel might range from 1 to more than 1000 per day; the implications of the corollary range in rates of fluid movement under unsaturated flow conditions are obvious.

Perhaps more significant than any of the preceding departures from an idealized physical setting, however, is the non-uniform nature of the porous materials at the many places where disposal operations might be practiced. The effect of this factor can be most readily demonstrated by a slight modification of the cross section shown in Figure 2. Such a revised section, in which clay lenses are present in the unsaturated zone, is shown in Figure 3. The permeability of the clay obviously is drastically less than that of the surrounding medium-grained sand. Thus, as fluid moves downward from the disposal pit, as shown by the arrows, its advance is unimpeded until it reaches the first clay lens,

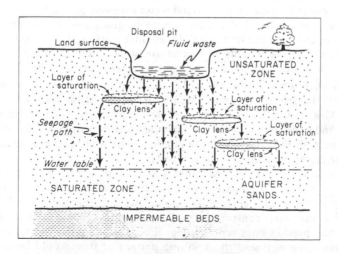

FIGURE 3. Cross section of disposal pit, showing seepage through an unsaturated zone occupied by clay lenses.

which represents a zone of greatly differing permeability. Subsequent events can be read as follows from the details of the sketch: the fluid collects on the upper surface of the clay lens, forming a small perched zone of saturation, and then spreads laterally until it finds a way around the lens; the fluid then continues downward, and the same cycle of events is repeated at each new clay lens encountered. When the fluid arrives at the water table, therefore, it may have traveled a significant distance laterally from the disposal pit. If very many clay lenses are present or if any lenses are of considerable areal extent, little imagination is required to appreciate the possible intricacies and the lateral spread of the fluid wastes as they travel to the water table.

The spreading of fluid waste, as depicted in Figure 3, can occur at any boundary between materials of substantially different permeabilities. The phenomenon is readily manifested when the difference in permeabilities is as little as one order of magnitude (10 times), and it matters not whether the fluid is endeavoring to cross a boundary into a region of lower or higher permeability. Predictions of where and how a fluid waste may travel from disposal site to the water table thus require detailed information on the physical characteristics, location, and extent of all pervious and impervious materials in the unsaturated zone.

Fluid Movement in the Saturated Zone

Inasmuch as the term ground water is properly applied only to water in the zone of saturation or that available to wells, the whole subject of ground water contamination ultimately must focus on the nature of fluid movement in that zone. In the preceding section the manner in which fluid waste might reach the water table was examined, and of interest now are some possible travel routes of the fluid once it has entered the ground water reservoir or aquifer.

A plan view of the idealized homogeneous ground water system shown in Figure 1 might be represented as shown in Figure 4. The contours indicate elevations of the water table in feet above some reference datum; the arrows generalize directions of ground water movement toward the stream. If a contaminant reaches the water table anywhere in the mapped region, it will move with the ground water in a fairly straight or definable path toward the stream. This would be the nature of the movement in ideal homogeneous systems under natural conditions, and therefore it would not be difficult to predict the path the contaminant might follow as well as its positions along the path after successively longer time intervals.

Usually, however, the natural flow system in an aquifer is modified at least locally by domestic, industrial, or municipal wells. In most of the region tributary to the stream (Figure 4), the arrows are essentially parallel to each other. If a well were constructed in this region and pumped at a steady rate, some of the parallel lines representing directions of ground water flow would bend toward the well and the resulting pattern of flow would be as shown in Figure 5. Skibitzke (3) has given a simple relation for expressing the width of the

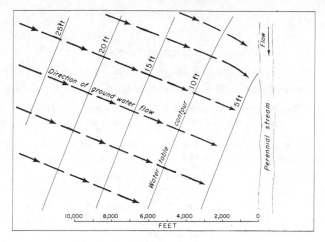

FIGURE 4. Water table map of an aquifer
discharging into a perennial stream.

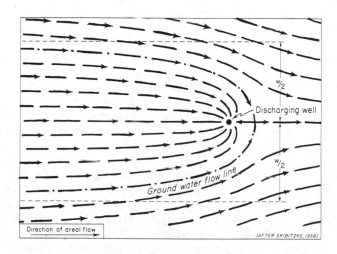

FIGURE 5. Flow lines near a discharging well constructed
in a region of parallel ground water flow.

area in a region of parallel flow within which the flow lines will converge upon
the well. The relation, for a unit thickness of the aquifer, may be given in the
form

$$w = \frac{Q}{PI} \qquad (2)$$

where

w = maximum width of the area of parallel flow within which the
      ground water will ultimately move to the well

Q = discharge rate per unit length of well bore

P = permeability of the saturated material comprising the aquifer

I = regional hydraulic gradient (head loss per unit length of flow path)
      under natural flow conditions before wells are introduced.

The dash-dot flow line in Figure 5 encloses the area of diversion of the regional
ground water flow to the well. If it is assumed that the natural hydraulic
gradient in the direction of ground water flow is 10 feet per mile, that the well
discharge per unit length of bore is 1 gallon per minute (or 1440 gallons per
day), and that the porous material of the aquifer is a uniform medium sand
having a permeability, under unit hydraulic gradient, of 70 feet per day, then, by
substitution in equation (2), it follows that

$$w = \frac{(1440/7.48)}{70(10/5280)} = 1450 \text{ feet}$$

This means that upgradient from the well a contaminant that reaches the water
table anywhere in the region whose maximum width is 1450 feet and whose
general shape is as shown in Figure 5 ultimately will emerge in the well
discharge. The converse of this example is also true, i.e., if the well were being
used to dispose of liquid waste at the rate of 1 gallon per minute, and the same
regional flow data were assumed, the contaminants would enter the aquifer and
move away from the well ultimately spreading to a band width of 1450 feet. In
either of the preceding examples, if the rate of well discharge or recharge were
doubled or tripled, the width w would also be doubled or tripled. Similarly, if
changes are postulated in either the permeability of the water-bearing material or
in the natural regional hydraulic gradient, the effect on w can be determined by
reference to equation (2).

Although the unsaturated zone is an important access route traveled by fluid
wastes from various types of disposal operations, ground water reservoirs or
aquifers are commonly in direct contact with surface water bodies such as
ponds, lakes, and streams that may contain contaminants. The aquifer shown in
Figures 1 and 4 is obviously in direct contact with a stream. As drawn, the
ground water system discharges into the stream and cannot be contaminated by
any wastes carried in the stream; however, situations occur in which a stream
feeds or recharges part of an aquifer. A fairly common situation is pictured in
Figure 6, which shows a supply well near a stream. As the well is pumped, the
water level will decline in the surrounding area, and the ultimate steady state
patterns of water table contours and ground water flow lines in the vicinity of
the well will be as shown in Figure 6. The patterns as drawn represent an
idealized homogeneous aquifer. If the stream shown in Figure 6 is polluted, the
water could readily move toward the pumped well and sooner or later appear in
the well discharge. Theis (4) developed equations for approximating the

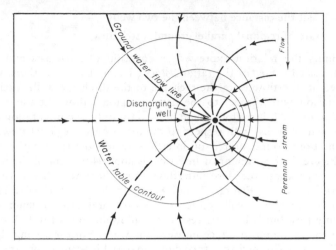

FIGURE 6. Flow lines and water table contours near a
discharging well constructed near a stream.

percentage of well discharge derived from the stream at different elapsed times
after pumping begins. Again, the converse of the preceding example is true, i.e.,
if the well were used to dispose of liquid waste, the contours in Figure 6 would
represent the built-up or mounded configuration of the water table and the
arrows denoting directions of ground water flow would be reversed.

Brief mention has been made heretofore of wells being used for the disposal
of liquid wastes. At a number of places in the east, wells have been used to
dispose of industrial wastes and storm runoff from city streets. This introduction
of contaminants via a well, directly into the saturated zone, prompts immediate
concern for the continued safe use of any nearby supply wells. Two examples
from a recent paper by daCosta and Bennett (2) afford some appreciation of the
factors controlling the possible interflow between a recharging well and a nearby
discharging well, in a region where parallel flow of ground water pre-existed.

The first example concerns a pair of discharging and recharging wells oriented
so that a line joining them is parallel to the direction of regional ground water
flow, and the recharging well is downgradient from the discharging well. The
ultimate steady state configurations of the water table contours and the flow
lines around such a pair of wells are shown in Figure 7. Within the shaded area,
flow lines that diverge from the recharging or disposal well subsequently
converge upon the discharging or supply well. The illustration was drawn under
the assumptions that the two wells have operated at the same rates over the same
period and that the following critical relation exists.

$$\frac{Q}{\pi a V_0} = 1.27 \qquad (3)$$

where

Q = rate of well discharge, or recharge, per unit length of well bore a   =   half the distance between the two wells

V_o =   rate of regional parallel ground water flow.

In determining the relation expressed as equation (3), daCosta and Bennett found that, for all values of the ratio $Q/\pi\, aV_o$ larger than 1.27, there would be some degree of interflow from the recharge to the discharge well, regardless of their orientation with respect to the natural regional flow. At the critical or limiting value of 1.27, interflow would be zero for only one orientation of a line joining the pair of wells, with respect to the direction of regional flow. For all other orientations, some degree of interflow would occur. For the situation shown in Figure 7, the amount of interflow is about 4 percent of the recharged fluid. The orientation for zero interflow (not illustrated) lies between the orientations shown in Figures 7 and 8. For values less than 1.27, some orientations of the wells will result in interflow and other orientations will ensure no interflow but will overly restrict the allowable rate of well discharge or recharge or the minimum distance between the wells. For the setting illustrated in Figure 7, if it is assumed that the aquifer material is again a uniform medium sand having a permeability of 70 feet per day, under unit hydraulic gradient, and that the regional hydraulic gradient or slope of the water table in the direction of flow is 10 feet per mile, the regional velocity of flow, V_o, is the product of 70 and (10/5280), or 0.13 foot per day. With this value substituted in equation (3), if follows that

$$\frac{Q}{a} = 0.52$$

Therefore, if the wells were 400 feet apart (a=200 ft), the rate of discharge or recharge would be slightly greater than 1/2 gallon per minute per foot of well

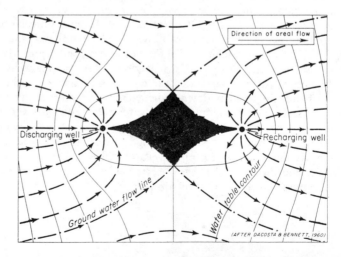

FIGURE 7. Flow lines and water table contours near a pair of recharging
and discharging wells aligned with the regional flow.

bore to develop the flow pattern shown in Figure 7. Other combinations of Q, a, and V_o that would satisfy equation (3) are readily made.

The second example is of a pair of discharging and recharging wells oriented so that a line joining them is at right angles to the direction of regional ground water flow. The ultimate steady state configurations of water table contours and flow lines around the wells are shown in Figure 8. The shaded area again encompasses the flow lines that diverge from the disposal well and subsequently converge upon the supply well. About 9 percent of the recharged (waste) fluid makes up this interflow. The illustration is drawn for the same conditions that were assumed for Figure 7; thus, the relations among Q, a, and V_o may be hypothesized and explored as before.

The preceding discussion of fluid movement in the saturated zone contains many idealizations and assumptions. Nevertheless with appropriate data on the location, extent, and physical properties of water-bearing materials and on the boundaries of the ground water flow system, it is possible to analyze the relative merits of a variety of waste disposal techniques and to describe the probable

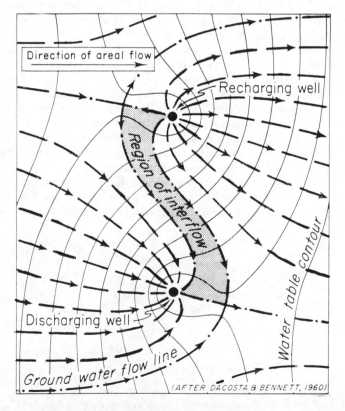

FIGURE 8. Flow lines and water table contours near a pair of recharging and discharging wells aligned at right angles to the regional flow.

consequences of each. Thus the hydrologist can contribute effectively to the design of disposal systems that will minimize or eliminate the danger of contaminating those parts of the ground water resource that are already being or may later be developed for beneficial use.

One important factor should be considered briefly. Many aquifers are composed of non-uniform kinds of material, such as sand or gravel. Thus, fine, medium, and coarse sands and gravels, as well as silts and clays, are often present in what may appear to be a meaningless arrangement of lenses and beds. To map the location and extent of each kind of material and to compute how fluid might move through it would be an endless task.

Fortunately, knowledge of erosional and depositional processes aids materially in filling in details between points where observations can feasibly be made. Furthermore, flow experiments with artificial aquifer models (Skibitzke, 1960, oral communication), simulating some of the non-uniform conditions found in nature, reveal that the flow regime is not chaotic. Figure 9 was drawn from a photograph of an aquifer-model experiment. The view simulates a broad expanse of aquifer with steady regional ground water flow from left to right, as shown. Medium sand comprises most of the aquifer, but trending through it are three continuous stringers of much coarser sand in somewhat sinusoidal paths. Where one stringer seems to disappear it simply dips below the aquifer-model surface to pass beneath the stringers that remain visible. Two different colored dyes, which could represent fluid wastes, are available at wells A and B, to be picked up and carried along by the flowing ground water. Different degrees of shading distinguish the courses taken by the fluids originating at A and B. Although the stringers of higher permeability afford some local preferential paths for flow, the regional movement of ground water sometimes crosses these stringers. Particularly significant, however, is the spreading of dyes or wastes as they traverse the region. The maximum spread appears to be about twice the amplitude of the sinusoidal, highly permeable stringers.

Only two dimensions of the flow system in the model aquifer appear in the illustration, but spreading similar to that shown occurs also in the third dimension. If representative flow paths could be sketched in perspective,

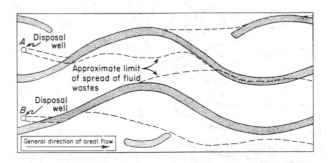

FIGURE 9. Plan view of a non-uniform aquifer, showing the spread of fluid wastes released into the regional ground water flow.

characteristic all three dimensions of the flow regime, it would be seen that considerable intertwining of the flow lines occurs. The implications are obvious with respect to predictions of travel paths for wastes originating at A and B. The hydrologist, therefore, must be able to recognize and describe those features of the local and regional geology that will most significantly affect the flow of ground water.

Hydrologic Factors

Up to this point in the discussion, a number of hydrologic factors have been covered inobtrusively. Their brief review, and mention of a few others will serve as a summary. The discussion has also been limited arbitrarily to environments of unconsolidated or granular porous media. Environments of consolidated rocks, such as granites, sandstones, and limestones, pose additional problems in defining the fluid-flow regimes that involve joint patterns, fracture patterns, solutional openings, and the rock structure.

Infiltration Rate

No attempt has been made to detail the mechanism of infiltration, i.e., the process by which fluids penetrate below the land surface. Common sense argues, however, that the finer the grain of the porous material, the slower the infiltration rate. A slow steady infiltration rate through the bottom of a disposal pit or through the porous material in which a tile drain field is laid could be the most significant insurance for delaying the arrival time of a fluid waste at some unwanted location.

Evaporation and Transpiration

The amount of fluid that might ultimately traverse the unsaturated zone is lessened by the processes of evaporation and transpiration. These processes, often combined into the single term evapotranspiration, account for the discharge to the atmosphere of large proportions of the fluid temporarily retained in the upper 8 or 10 feet of the unsaturated zone.

Unsaturated Flow

Rigorous mathematical description of fluid movement through the unsaturated zone, where the pore spaces are only partly filled with fluid, is difficult. The movement obviously is strongly related to the rate and manner in which the fluid is first introduced to the zone. Other factors include the amount and geometry of pore space in the porous material, the magnitude and direction of temperature and chemical gradients, and such fluid properties as density, viscosity, and surface tension.

Saturated Flow

Fluid movement in the saturated zone, where the pore spaces are filled with fluid, has been described in many scientific papers and reports. Analysis of the flow regime requires knowledge of the geometry of the ground water system and how it is connected with surface water bodies or sources of recharge, the nature of the porous material with respect to fluid movement through it, and the head distribution.

Permeability

Different kinds of rocks and earth materials resist, to differing degrees, the movement of fluids through them. The range in permeabilities from the tightest clays to the coarsest gravels, in terms of the kind of velocity units used in this paper, exceeds nine orders of magnitude. In many earth materials there are significant variations in permeability with distance and direction. Especially important is the fact that permeability in the vertical direction is commonly much less than in the horizontal direction. This is to be compared with predominantly vertical movement of fluid in the unsaturated zone and horizontal movement in the saturated zone.

Non-Uniformity of Porous Media

Flow in both the unsaturated and the saturated zones can be greatly affected not only by changes in texture within a given kind of porous material but also by the presence in that material of lenses or beds of other kinds of porous material. In alluvial valleys and in areas that were once glaciated, the distribution of different kinds of porous media is random. The analysis of problems of ground water contamination in such environments requires the exercise of the best hydrologic skills.

Conclusion

Only a few principles of fluid movement in porous media and a few ground water flow systems of simple geometry have been covered in this brief paper. Many ramifications and extensions can be found in the voluminous literature on the occurrence and movement of ground water. The cited references in particular define segments of the science of ground water hydrology that warrant careful study prior to analysis of situations of actual or potential ground water contamination.

The consequences of ground water contamination can be just as damaging to water users as the pollution of surface streams. In fact it can be argued that the consequences are far more damaging because they persist over much longer periods of time after the contaminating source has been eliminated. It would appear prudent, therefore, to guard against contamination of the ground water resource in the first instance, rather than to engage in long expensive rehabilitation measures after the damage has been done.

References

1. Baver, L. D. 1948. Soil physics. New York: John Wiley & Sons, Ch. VI.
2. daCosta, J. A., and R. R. Bennett. 1961. The pattern of flow in the vicinity of a recharging and discharging pair of wells in an aquifer having areal parallel flow. *In* Commission of Subterranean Waters. Internat. Union Geodesy and Geophysics, Internat. Assoc. Sci. Hydrology, General Assembly at Helsinki, 1960, pub. 52, p. 524-536.

3. Skibitzke, H. E. 1958. The use of radioactive tracers in hydrologic field studies of ground-water motion. Internat. Union Geodesy and Geophysics, Internat. Assoc. Sci. Hydrology, General Assembly at Toronto, 1957, 2:243-252.
4. Theis, C. V. 1941. The effect of a well on the flow of a nearby stream. Trans. Amer. Geophysical Union, part 3, p. 734-738.
5. Walton, G. 1960. ABS contamination. Amer. Water Works Assoc. 52(11)/.
6. Wyllie, M. R. J., and G. H. F. Gardner. 1958. The generalized Kozeny-Carman equation: Its application to problems of multiphase flow in porous media, part 2. World Oil, Production Sect. 146:210-228.

Monitoring of Changes in Quality of Ground Water

H. E. LeGrand*

Introduction

Ground water of acceptable quality is available in most areas except where an arid climate and relatively impermeable rocks combine to eliminate or reduce its occurrence. It is commonly in contact with water of inferior quality; the water of inferior quality may be naturally occurring salty water that commonly underlies the fresh water, or it may be enclaves of contaminated water from man's wastes that lie in the fresh-water bodies. Such actions as disposal of wastes on and in the ground and pumping of water from wells cause a dispersion of contaminated water; the movement of contaminated water may naturally be toward them as a result of the depressed water level near the cone of depression. Simple and economic methods of determining precisely the boundaries between contaminated and uncontaminated water are not available. Yet, the need to know the approximate positions of many of these boundaries is essential in order to prevent or reduce actual contamination of water supplies and to plan for future water supplies and waste disposal sites. In the absence of indirect methods, much reliance is placed on monitoring wells. A monitoring well may be defined as one strategically located near potentially mobile contaminated water, the purpose of the well being to allow a forewarning of the spread of objectionable contamination. A test well is designed to determine the geologic and hydrologic conditions at a particular place; it becomes a monitoring well if it is used to detect hydrologic changes with time.

An objective of monitoring wells may be to determine whether contaminated water will move to certain well sites or to determine the direction and extent of movement of contaminated water. The need for monitoring a particular area is dependent on the probability of contamination and the seriousness of contamination. There may be a need also for verification—to substantiate or test the inferences of persons who evaluate the underground conditions and to prove

Reprinted from *Journal of Ground Water*, Vol. 6, no. 3, 1968, pp. 14-18, with permission of the author and The National Water Well Association.

*Water Resources Division, U.S. Geological Survey, Raleigh, North Carolina.

or disprove the presence or absence of contamination at a particular place. In the latter case the use of tracers in ground water may be attempted.

In spite of the seemingly strong justification for monitoring, there are many discouraging aspects that are considered in this paper. In one context the discouraging aspects involve costliness, poor timing or slowness in getting satisfactory answers, and indefinite planning for water supplies or waste disposal projects.

Movement of Contaminants in the Ground

Haphazard plans for monitoring wells are almost certain to result in excessive costs and to fail in their objectives. An early consideration for a monitoring program is the hydrogeologic framework that relates to the movement of water and of contaminants that might be with it. Brief, useful generalizations about the geologic framework are not altogether satisfactory, but we should be constantly aware that ground water flows under the influence of gravity and difference in pressures from points of higher potential to points of lower potential; to the extent that it can take preferred paths, water tends to flow through permeable materials and around relatively impermeable materials. The frequency of precipitation in humid regions is sufficient to keep the water table relatively close to the ground surface, and the consequent mounding of water beneath interstream areas causes a continuous subsurface flow of water to nearby perennial streams. Thus, even the uninitiated in hydrology can get a general idea of the gross direction of movement of ground water in humid regions. In arid regions, however, the areas of natural ground-water discharge are more widely scattered, and the general movement of water may be less discernible. In arid regions some stretches of most streams are influent—that is, water and waste from the stream may seep into the ground; yet, this is not the case with the generally effluent type of streams in most humid regions under natural conditions, although pumping of wells near a stream may reverse the natural flow.

Some questions that might be asked at an early stage in order to define the hydrogeologic setting are: Does water move in porous granular material or in fracture or solution openings in consolidated rock? What is the general distribution of permeable and relatively impermeable materials? Is the direction of natural movement of water to some discharge area approximately known? To what extent is the natural movement of water diverted by pumping of wells or to what extent might it be diverted in the future?

The fact that contaminants move beneath the ground in the same direction as water is of some help but does not simplify the understanding of contaminant movement. Some contaminants, such as chlorides and certain minor elements, appear to move at about the same rate as the entraining water and attenuate, or decrease in potency, only by dilution. Other contaminants seem not to move as fast as the water, and in fact they may be retained near their source because of sorption on earth materials, or they may tend to die away by some intrinsic mechanism; these habits of attenuation by some contaminants result in what appears to be a slow rate of movement. Where contamination is continuous at a source the movement of contaminated water will likely extend in the direction of water movement at a diminishing rate; where contamination is discontinuous, as a single release of waste, the movement of contaminated water will be similar to that from a continuous source, but after the maximum distance of contaminant movement has been reached there will be a regression in movement as the contaminated zone shrinks in size.

Planning a Monitoring Program

Before a monitoring program is undertaken, some preliminary evaluation should be made of (1) the pertinent features of the hydrogeologic framework, (2) characteristics of the particular contaminants as to attenuation and movement in the ground environment, and (3) man-made contingencies causing movement of contaminated water, whether from pumping of wells, disposal of wastes, or accidents that could cause contamination. The preliminary evaluation should predicate the possibility of contamination.

A desirable goal, seldom attained, is to know the degree of contamination at many points from the source of contamination or to have a map showing the distribution of the contaminated water zone and uncontaminated water zone and maps showing prior distribution of these zones. The extent of monitoring necessary to realize this goal is commonly too great to be practical. Therefore, in the interest of economy it is necessary to limit the monitoring program to a minimum that will still give satisfactory answers. In many cases the position of a strategic segment of the boundary between uncontaminated and contaminated water and the degree of movement of this boundary will suffice. Plans for monitoring should consider such aspects as approximate geometry of the contaminated and uncontaminated zones, the potential for advancement of a contaminated front toward existing and potential water supplies, and remedial measures if contamination is likely.

The quality of water, from the places where uncontaminated water occurs to the places where highly contaminated water occurs, may conveniently be divided into four categories or gradational zones as follows:

    A.   Zone of high concentration of contaminants

    B.   Contamination slight to moderate and water objectionable for use (later precursor zone)

    C.   Contamination detectable but quality of water not seriously objectionable (early precursor zone)

    D.   Uncontaminated zone

An early objective is to determine the approximate breadth of zones B and C and the trend of movement of these zones. If the concern is of lateral encroachment, the breadth of zone B or C may range from a few feet to a few hundred feet for contaminants from waste sites and may range from a few hundred feet to a few miles for naturally occurring mineralized water. If the concern is of vertical encroachment, the breadth of zone B or C may be a few feet to several tens of feet; the progression of encroachment in the vertical field may be retarded or prevented by the occurrence of beds of widely differing permeability.

The four gradational zones may be adapted to particular situations where contaminated water has the potential for moving toward uncontaminated water. Prior to monitoring, estimates must be made in each case as to the degree of contamination between uncontaminated and highly contaminated water, even though the estimate may be crude. Following an early estimate, priority may be given to placing one or more wells in what is considered to be the early precursor zone. The earliest aspect of contamination may be detected by small quantities of the contaminant or by some constituent that travels in the water at a faster rate than the contaminant and that may be a forewarner of the contaminant. For example, in cases of lateral salt-water encroachment a water having a few parts per million chloride more than that in the uncontaminated zone would be expected in the early precursor zone. Where contaminants from waste sites undergo attenuation by sorption or some decay mechanism, the contaminants may either never reach the early precursor zone or may reach it long after the water that originally carried it reaches the zone; in some cases, such as those involving sewage wastes, an increase in chloride or hardness content of the water would be the early precursor.

In view of the high costs that may develop in a monitoring program, it is desirable to get optimum value from each monitor well. It is unlikely that only one monitor well will suffice unless there is good assurance that it will be effective or if the consequences of it being ineffective are slight. In almost every case some information will be available concerning the approximate direction of movement of the contaminated water, and inferences from the information, however meager it may be, will direct the position of the first well or wells. Perhaps an early approach would be in many cases to have 2, 3, or more wells alined perpendicular to the inferred direction of contaminant movement near the boundary of zone D and zone C as in Figure 1. If slight or moderate contamination is detected, plans then may be made to put in 2 or more wells somewhat nearer the water supply and somewhat offset from the earlier monitor wells. Several types of situations are shown in Figure 1 indicating how contaminated water moves from a source of contamination. The 3 wells shown in each situation are favorably located to get optimum value, but rarely would all 3 wells be so favorably located. The reader can readily see the high probability of having one or more poorly located wells. In addition to its location, a good monitoring well must be of proper depth. Where contaminated water from waste sites occurs, it commonly tends to be more concentrated in the upper part of the zone of saturation. Where naturally occurring salty water occurs it tends to be below water of better quality. Figure 2 shows favorable positioning of test and monitor wells to determine the general boundary between fresh and salty water and to monitor the movement of this boundary. Each monitoring program must rest on its own merits. In order to minimize the number of wells, it is essential at all stages to use all pertinent data and draw the

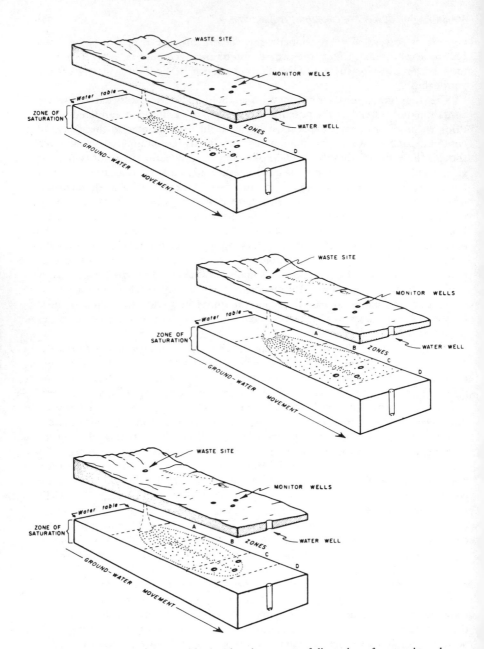

FIGURE 1. View of water table showing three types of dispersion of contaminated water from a waste site. Degree of contamination in direction of ground-water gradient is shown by zones, ranging from heavily contaminated zone (A) to uncontaminated zone (D). Favorable locations of 3 monitoring wells are shown for each type, but seldom would the first 3 monitoring wells be so favorably located.

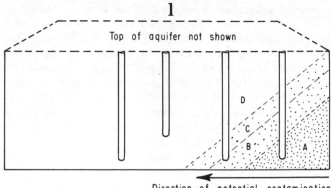

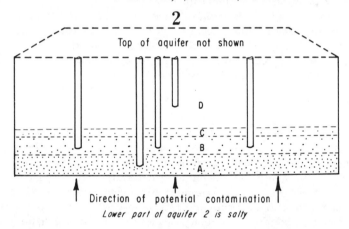

FIGURE 2. Proile view of aquifers (1 and 2) showing fresh water in contact with salty water. Heavily contaminated zone (A) is separated from fresh-water zone (D) by moderately contaminated zone (B) and slightly contaminated zone (C). Test and monitor wells must locate approximate boundary between fresh and salty water and should help to determine extent of encroachment of salty water.

best inferences even though the inferences may be no more than "educated guesses."

Tracers

Much consideration has been given to the use of tracers, such as dyes, certain chemical constituents, and radioactive isotopes, to determine the direction and rates of water movement in the ground.

Tracers in ground water have limited value, but their use should not be discounted without consideration. There is difficulty in selecting a correct tracer that will reflect the movement of water and (1) that will move at the same rate as water, (2) that will not attenuate except by dilution, (3) that is inexpensive, (4) that can be injected easily, and (5) that will be unobjectionable as to its own zone of influence. Some selected materials to be put in water as tracers tend to become a part of the water and to move directly with the water; examples of these are chlorides and tritium. Other materials, including some dyes and many cations, tend to be sorbed on clays and other earth materials and thus may not extend far from places of tracer injection.

If a problem concerns contaminants that move slower than water—that is, they attenuate by decay and (or) sorption—the tracer selected would not likely reveal the rate of movement of the contaminant. Another difficulty is that results of tracer tests commonly take too long for the purpose, except where water moves rapidly through tubular openings such as occur in some limestone formations. The time for completion of tracer tests cannot be easily shortened. A short segment of travel of the tracer may not be representative because neither the gradient nor the time of travel is necessarily uniform along the path of movement. Owing to great ranges in hydrogeologic conditions, the results of tracer tests may have only limited transfer value to a problem area.

Summary

The movement of contaminated water in the ground may have no surface manifestation, and the extent of movement of the contaminated water is indeterminate except in contexts of probability or degree of contamination. As complex as the movement of uncontaminated water may appear to be, the complexity of the movement of contaminated water is compounded because of the tendency for the entrained contaminants to be attenuated to some degree by (1) dispersion and dilution, (2) some die-away or decay mechanism, or (3) sorption on earth materials. The complex interrelated factors prevent any simple or precise solution or index to contaminant movement in the ground. In view of the widespread distribution of contaminated water, the need for information obtained by test and monitoring wells is determined by the adequacy of information obtained by other ways, by the probability of contamination at certain strategic places, and by the consequences of contamination.

Obtaining needed information by simple means commonly falls short of expectations. Thus, there is frequently a need for test wells to delineate critical segments of contaminated zones and a need for monitoring wells to determine whether contaminated water is likely to reach places where it is not acceptable. Seldom will sufficient money and time be available to completely delineate and monitor a contaminated zone. One well may help if there is not complete reliance on the well; if located strategically two or three wells may help, but complete adequacy of well data may seldom be realized.

The following questions, among others, need to be asked at an early stage of monitor planning. What is the probability or possibility of contamination at the current state of knowledge? What are consequences of contamination? Does the decision depend on a clear-cut distinction between acceptability or unacceptability? Does the decision depend on gradational degrees of acceptability? What are the consequences of a poor test or monitoring well? Is the information merely (1) of no importance, (2) of slight importance, or (3) possibly misleading? How far can each inference go without excessive danger of being misleading? Even if contaminated water is moving toward a well, what conditions could prevent contaminated water from reaching it? To what extent can test and monitoring wells be planned? What is the present plight and what is the outlook for the future as to the need for monitoring and the success of monitoring?

Technology of evaluating the behavior of contaminants in the ground is at a relatively low stage. Consequences of this weakness in technology are the contamination of ground-water supplies in some cases and excessive costs and inconveniences in many cases. Ignorance shrouds and obstructs the proper planning of the development of ground-water supplies and of waste disposal. Such projects need to be considered in a framework, the essential parts of which include economics, maintenance of acceptable health practices, reasonableness, and the common good. The present situation is poor, and no improvement is likely in the immediate future. An overall view suggests that an increase in pumping of wells, an increase in waste to be disposed, and other actions by man will more frequently bring contaminated subsurface water to places where it is not acceptable. Decrees by a State Water Department or other government agencies are a common consideration for coping with the complex problems, but decrees generally are indiscriminate to the extent that undue restrictions and hardships are placed on too many people. Indiscriminate monitoring of contaminated water is also not a simple solution because of undue costs. The need for monitoring will increase in the future, of course. Yet, the proper objective is to improve the technology of determining the distribution of contaminated water in the ground so that monitoring can be minimized. As a prerequisite to monitoring it is helpful to have a synthetic hydrogeologic framework in which the behavior of the contaminated water is conceived. Such a conceptual model, using pertinent data that are expeditiously available will help to assess the need for monitoring and to conduct a monitoring program with optimum results where it is necessary.

References

1. Deutsch, Morris. 1963. Ground-water contamination and legal controls in Michigan. U.S. Department of the Interior, Geol. Survey Water-Supply Paper 1691.
2. LeGrand, H. E. 1965. Patterns of contaminated zones of water in the ground. Water Resources Research 1:83-95.
3. McGuinness, C. L. 1963. The role of ground water in the national water situation. U.S. Department of the Interior, Geol. Survey Water-Supply Paper 1800.
4. State Water Pollution Control Board. 1954. Report on the investigation of travel of pollution. State Water Pollution Control Board, Sacramento, Calif., Pub. 11.
5. Stiles, C. W., H. R. Crohurst, and G. E. Thompson. 1927. Experimental bacterial and chemical pollution of wells via ground water, and the factors involved. U.S. Department of Health, Education, and Welfare, Public Health Service Hygienic Lab. Bull. 147, 136 p.
6. U.S. Department of Health, Education, and Welfare, Public Health Service. 1961. Proceedings of 1961 symposium on ground-water contamination. Public Health Service Tech. Rept. W61-5.

Hydrologic Implications of Solid-Waste Disposal

William J. Schneider*

Introduction

The disposal of solid-waste material—principally garbage and rubbish—is primarily an urban problem. However, unlike liquid waste disposal of sewage and industrial effluents, the problem has received only limited recognition. It is common practice in many metropolitan areas to overlook or ignore the consequences of waste-disposal programs. The full scope of the problem, though, cannot be ignored.

The urban population of the United States is now producing an estimated 1,400 million pounds of solid wastes each day. Disposal of these wastes is a major problem of all cities. In many instances, seemingly endless streams of trucks and railroad cars haul these wastes long distances—as much as hundreds of miles—to disposal sites. Based on a volume estimate of 5.7 cubic yards per ton of waste, this refuse is sufficient to cover more than 400 acres of land per day to a depth of 10 feet. Local governments spend an estimated $3 billion each year on collection and disposal, a sum exceeded in local budgets only by expenditures for schools and roads.

The disposal of these solid wastes poses many problems to local government agencies. Unfortunately this problem is handled by many governments on the basis of expediency without due regard to environmental considerations. Garbage and rubbish are collected, hauled minimum distances commensurate with public acceptance, and dumped. Occasionally, the waste is either burned or mixed with soil to provide landfill. As long as the procedure removes the refuse and as long as the disposal site is not a health hazard and does not offend esthetic values too greatly, the operation is considered successful. Overlooked or even ignored is the effect of the disposal on the total environment, including the water resources of the area. Although the disposal of solid wastes can create many serious health, esthetic, and environmental problems, only the hydrologic implications—the effect upon water resources—are considered in this report.

Reprinted from *U.S. Geological Survey Circular 601-F*, 1970, pp. F1-F10.
*Water Resources Division, U.S. Geological Survey, Washington, D.C.

Types of Solid Wastes

Our urban society generates many types of solid wastes. Each may exert a different influence on the water resources of an area. In order to understand the effect of each type, it is necessary to identify the various types as to the principal constituents. Table 1 lists the various categories and sources of refuse material primarily generated by urban activities. Not included are wastes from industries and processing plants; hazardous, pathological, or radioactive wastes from institutions and industries; solids and sludge from sewage-treatment plants; and other special types of solid wastes. These items usually pose special handling problems and are usually not a part of normal municipal solid-waste-disposal programs. The following descriptions of the categories of solid wastes are abbreviated from descriptions by the American Public Works Association (1966).

Waste refers to useless, unused, unwanted, or discarded materials including solids, liquids, and gases.

Refuse refers to solid wastes which can be classified in several different ways. One of the most useful classifications is based on the kinds of material: garbage, rubbish, ashes, street refuse, dead animals, abandoned automobiles, industrial wastes, demolition wastes, construction wastes, sewage solids, and hazardous and special wastes.

Garbage is the animal and vegetable waste resulting from the handling, preparation, and cooking of foods. It is composed largely of putrescible organic matter and its natural moisture. It originates primarily in home kitchens, stores, markets, restaurants, and other places where food is stored, prepared, or served.

Rubbish consists of both combustible and noncombustible solid wastes from homes, stores, and institutions. Combustible rubbish is the organic component of refuse and consists of a wide variety of matter that includes paper, rags, cartons, boxes, wood, furniture, bedding, rubber, plastics, leather, tree branches, and lawn trimmings. Noncombustible rubbish is the inorganic component of refuse and consists of tin cans, heavy metal, mineral matter, glass, crockery, metal furniture, and similar materials.

Ashes are the residue from wood, coke, coal, and other combustible materials burned in homes, stores, institutions, and other establishments for heating, cooking, and disposing of other combustible materials.

Street refuse is material picked up by manual and mechanical sweeping of streets and sidewalks and is the debris from public litter receptacles. It includes paper, dirt, leaves, and other similar materials.

Dead animals are those that die naturally or from disease or are accidentally killed. Not included in this category are condemned animals or parts of animals from slaughterhouses which are normally considered as industrial waste matter.

Abandoned vehicles include passenger automobiles, trucks, and trailers that are no longer useful and have been left on city streets and in other public places.

TABLE 1. Classification of refuse materials.
[Adapted from American Public Works Association (1966)]

Kind of refuse	Composition	Source
Garbage	Wastes from preparation, cooking, and serving of food; market wastes; wastes from handling, storage, and sale of produce.	Households, restaurants, institutions, stores, and markets.
Rubbish	Combustible: paper, cartons, boxes, barrels, wood, excelsior, tree branches, yard trimmings, wood furniture, bedding, and dunnage. Noncombustible: metals, tin cans, metal furniture, dirt, glass, crockery, and minerals.	Do.
Ashes	Residue from fires used for cooking and heating and from onsite incineration.	Do.
Trash from streets	Sweepings, dirt, leaves, catch-basin dirt, and contents of litter receptacles.	Streets, sidewalks, alleys, vacant lots.
Dead animals	Cats, dogs, horses, and cows	Do.
Abandoned vehicles	Unwanted cars and trucks left on public property	Do.
Demolition wastes	Lumber, pipes, brick, masonry, and other construction materials from razed buildings and other structures.	Demolition sites to be used for new buildings, renewal projects, and expressways.
Construction wastes	Scrap lumber, pipe, and other construction materials	New construction and remodeling.

Methods of Solid-Waste Disposal

The disposal of these solid wastes generated by our urban environment is generally accomplished by one or more of six methods. All are currently in use to one degree or another in various parts of the United States. To a large extent, the method of waste disposal in any particular area depends upon local conditions and, to some extent, upon public attitude. In many areas several methods are employed. Each has its unique relation to the water resources of the area. The six general methods of solid waste disposal are:

1. Open dumps.
2. Sanitary landfill.
3. Incineration.
4. Onsite disposal.
5. Feeding of garbage to swine.
6. Composting.

Open dumps. Open dumps are by far the oldest and most prevalent method of disposing of solid wastes. In a recent survey, 371 cities out of 1,118 surveyed stated that this method was emphasized within their jurisdictions. In many cases, the dump sites are located indiscriminately wherever land can be obtained for this purpose. Practices at open dumps differ. In some dumps, the refuse is periodically leveled and compacted; in other dumps the refuse is piled as high as equipment will permit. At some sites, the solid wastes are ignited and allowed to burn to reduce volume. In general, though, little effort is expended to prevent the nuisance and health hazards that frequently accompany open dumps.

Sanitary landfill. As early as 1904, garbage was buried to provide landfill. Although in subsequent years, the practice was used by many cities, the technique of sanitary landfill as we know it today did not emerge until the late 1930's. By 1945, almost 100 cities were using the practice, and by 1960 more than 1,400 cities were disposing of their solid wastes by this method.

Sanitary landfill consists of alternate layers of compacted refuse and soil. Each day the refuse is deposited, compacted, and covered with a layer of soil. Two types of sanitary landfill are common: area landfill on essentially flat land sites, and depression landfill in natural or manmade ravines, gulleys, or pits. Depth of the landfill depends largely on local conditions, types of equipment, availability of land, and other such factors, but it commonly ranges from about 7 feet to as much as 40 feet as practiced by New York City.

In normal operation, the refuse is deposited and compacted and covered with a minimum of 6 inches of compacted soil at the end of each working period or more frequently, depending upon the depth of refuse compacted. Normally about a 1:4 cover ratio is satisfactory; that is, 1 foot of soil cover for each 4-foot layer of compacted refuse. Ratios as high as 1:8, however, have been used. The final cover is at least 2 feet of compacted soil to prevent the problems associated with open dumps.

Incineration. Incineration is the process of reducing combustible wastes to inert residue by burning at high temperatures of about 1,700° to 1,800°F. At these temperatures all combustible materials are consumed, leaving a residue of ash and noncombustibles having a volume of 5 to 25 percent of the original volume.

Although incineration greatly reduces the volume and changes the material to inorganic matter, the problem of disposal is still present. Much of the residue is hauled to disposal sites or is used for landfill, although the land required for disposal of the residue is about one-third to one-half of that required for sanitary landfill. Some cities require that combustible materials be separated from noncombustibles prior to collection, while others use magnetic devices to extract ferrous metal for salvage.

The combination of urban growth, increasing per capita output of refuse, and the rising costs of land for sanitary landfills has stimulated the use of incineration for solid-waste disposal. Today, there are an estimated 600 central-incinerator plants in the United States with a total capacity of about 150,000 tons per day.

Onsite disposal. With the increasing rate of production of solid wastes in the urban environment, there is a growing trend toward handling this waste in the home, apartment, and institution. Onsite disposal has become increasingly popular during the past decade as a way of minimizing the waste problem at its source. Most widely used devices for onsite disposal are incinerators and garbage grinders.

Onsite incineration is used widely in apartment houses and institutions. The incinerators do, however, require constant attention to insure proper operation and complete combustion. Domestic incinerators for use in individual homes are not a major factor in solid-waste disposal, nor are they likely to be a major factor in the near future. Maintenance and operating problems are usually considerable.

Garbage grinders, on the other hand, are becoming increasingly prevalent in homes for disposal of kitchen food wastes. It is estimated that more than a million grinders are now in home use. The grinders are installed in the waste pipe from the kitchen sink; food wastes are simply scraped into the grinder, the grinder is started, and the water turned on. The garbage is ground and flushed into the sanitary-sewer system. In some local communities, garbage grinders have been installed in every residence as required by local ordinance.

Swine feed. The feeding of garbage to swine has been an accepted way of disposing of the garbage part of solid wastes from urban areas for quite some time. Even as late as 1960, this method was employed in 110 American cities out of 1,118 cities surveyed on their solid-waste-disposal practices. In addition to the municipal practices of using garbage for swine feed, many cities and municipalities permit private haulers to service restaurants and institutions to collect garbage for swine feed. The feeding of raw garbage led to a wide-spread virus disease in the middle 1950's, which affected more than 400,000 swine. As a result, all States now require that garbage be cooked before feeding to destroy contaminating bacteria and viruses. However, according to the American Public

Works Association (1966), more than 10,000 tons of food wastes—about 25 percent of the total quantity of garbage produced—is still used daily in the United States as swine feed.

Composting. Composting is the biochemical decomposition of organic materials to a humuslike material. As practiced for solid-waste disposal, it is the rapid but partial decomposition of the moist, solid-organic matter by aerobic organisms under controlled conditions. The end product is useful as a soil conditioner and fertilizer. The process is normally carried out in mechanical digesters.

Although popular in Europe and Asia where intensive farming creates a demand for the compost, the method is not used widely in the United States at this time. Composting of solid-organic wastes is not practiced on a full-scale basis in any large city today. Although there are several pilot plants in operation, it does not seem likely that composting will be a major method of solid-waste disposal.

The selection of one or more of these methods of solid-waste disposal by a municipality depends largely on the character of the municipality. Geographic location, climate, standard of living, population distribution, and public attitudes play important roles in the selection. In general, the natural resources and environmental factors have been given only small recognition in this selection. Only recently has there been a considerable upsurge of scientific interest in the effects of solid-waste disposal on our water resources.

Hydrologic Implications

Types of Pollution

The disposition of solid wastes in open areas carries with it an inherent potential for pollution of water resources, regardless of the manner of disposal or the composition of the waste material. Of the six principal methods of solid-waste disposal, only swine feeding and composting offer no direct possibility of pollution of water resources from the waste material itself. Quite the contrary: properly composted garbage is a soil conditioner that improves the permeability of the soil and may actually assist in improving the quality of water that percolates through it. Although the cooked garbage that is fed to swine does not directly contribute to pollution of water resources, the manure from the feedlots may cause serious problems if not managed properly.

The type of pollution that may arise is directly related to the type of refuse and the manner of disposal. Leachates from open dumps and sanitary landfill usually contain both biological and chemical constituents. Organic matter, decomposing under aerobic conditions, produces carbon dioxide which combines with the leaching water to form carbonic acid. This, in turn, acts upon metals in the refuse and upon calcareous materials in the soil and rocks, resulting in increasing hardness of the water. Under aerobic conditions, bacterial action decomposes organic refuse, releasing ammonia, which is ultimately oxidized to form nitrate. In both landfills and open dumps, where decomposition is

TABLE 2. Percentages of materials leached from refuse and ash, based on weight of refuse as received.
[Adapted from Hughes (1967)]

Material leached	Percentage leached under given conditions *					
	1	2	3	4	5	6
Permanganate value ___ 30 min	0.039					
Do ___ 4 hr	.060	.037				
Chloride	.105	.127		0.11	0.087	
Ammonia nitrogen	.055	.037		.036		
Biochemical oxygen demand	.515	.249		1.27		
Organic carbon	.285	.163				
Sulfate	.130	.084		.011	.22	0.30
Sulfide	.011					
Albumin nitrogen	.005					
Alkalinity (as CaCO₃)				.39	.042	
Calcium				.08	.021	2.57
Magnesium				.015	.014	.24
Sodium			.260	.075	.078	.29
Potassium			.135	.09	.049	.38
Total iron				.01		
Inorganic phosphate				.0007		
Nitrate					.0025	
Organic nitrogen	.0075	.0072		.016		

*Conditions of leaching:
1. Analyses of leachate from domestic refuse deposited in standing water.
2. Analyses of leachate from domestic refuse deposited in unsaturated environment and leached only by natural precipitation.
3. Material leached in laboratory before and after ignition.
4. Domestic refuse leached by water in a test bin.
5. Leaching of incinerator ash in a test bin by water.
6. Leaching of incinerator ash in a test bin by acid.

accomplished by bacterial action, the leachate has a high biochemical oxygen demand (BOD).

Table 2 indicates the magnitude of the constituents leached from solid wastes under various conditions. These data were compiled by Hughes (1967) from various sources.

Relation to Hydrologic Regimen

That part of the hydrologic regimen associated with pollution from solid-waste disposal begins with precipitation reaching the land surface and ends with the water reaching streams from either overland or subsurface flow. The manner in which this precipitation moves through this part of the cycle determines whether or not the water resource will become polluted.

Precipitation on the refuse-disposal site will either infiltrate the refuse or run off as overland flow. In open dumps, there is little likelihood of direct runoff unless the refuse is highly compacted. In sanitary landfills, the rate of infiltration is governed by the permeability and infiltration capacity of the soil used as cover for the refuse. A part of the water entering the refuse percolates downward to the soil zone and eventually to the water table. If the water table is above the bottom of the refuse deposit, the percolating water travels only vertically through the refuse to the water table. During the vertical-percolation process the water leaches both organic and inorganic constituents from the refuse.

Upon reaching the water table, the leachate becomes part of and moves with the ground-water flow system. As part of this flow system, the leachate may move laterally in the direction of the water-table slope to a point of discharge at the land surface. In general, the slope of the water table is in the same direction as the slope of the land. The generalized movement of leachate in this part of the hydrologic cycle is shown in Figure 1.

There are several well-documented cases of pollution caused by leachates from solid-waste-disposal sites, especially those compiled by the California Water Pollution Control Board (1961). Most of these studies, however, were able to determine only that the pollution originated from solid-waste-disposal sites; few, if any, data on the gross magnitude of the pollution and its fate in the hydrologic cycle are available.

One well-documented case is that of pollution from about 650,000 cubic yards of refuse deposited in a garbage dump near Krefield, Germany, over a 15-year period in the early 1900's. High salt concentrations and hardness were detected in ground water about a mile downgradient from the site within 10 years of operation. Concentrations up to 260 mg/l (milligrams per liter) of chloride and a hardness of 900 mg/l were measured—an increase of more than sixfold in chloride concentration and fourfold in hardness. The pattern of pumping of wells in the area precludes detailed understanding of the course of the pollution in the ground water, but wells near the dumping site were still contaminated 18 years later.

In Schirrhof, Germany, ashes and refuse dumped into an empty pit extending

FIGURE 1. Generalized movement of leachate through the land phase of the hydrologic cycle.

below the water table resulted in contamination of wells about 2,000 feet downstream. The contamination occurred 15 years after the dump was covered; measures of hardness up to 1,150 mg/l were recorded as compared with 200 mg/l prior to the contamination.

In Surrey County, England, household refuse dumped into gravel pits polluted the ground water in the vicinity. Refuse was dumped directly into the 20-foot-deep pits where water depth averaged about 12 feet. Maximum rate of dumping was about 100,000 tons per year over a 6-year period, and this occurred during the latter part of the period of use (1954-60). Limited observations on water quality extending less than a year after the closing of the pits showed chloride concentrations ranging from 800 mg/l at the dump site, through 290 mg/l in downgradient adjacent gravel pits, to 70 mg/l in pits 3,500 feet away. Organic and bacterial pollution were detected within half a mile of the dumping sites, but not beyond. Because of the limited study period and the slow travel of the pollutants, the maximum extent of pollution was not determined.

More recently, a study was made of the ground-water quality associated with four sanitary landfill sites in northeastern Illinois (Hughes and others, 1969). At the DuPage County site, total solids of more than 12,500 mg/l and chloride contents of more than 2,250 mg/l were measured in samples collected about 20 feet below land surface under the fill. These were by far the highest concentrations measured at any of the four sites. In general, total solids ranged from 2,000-3,000 mg/l under the fill to as low as 223 mg/l adjacent to the fill.

Hydrologic Controls

The movement of leachate from a waste-disposal site is governed by the physical environment. Where the wastes are above the water table, both chemical and biological contaminants in the leachate move vertically through the zone of aeration at a rate dependent in part upon the properties of the soils. The chemical contaminants, being in solution, generally tend to travel faster than biological contaminants. Sandy or silty soils especially retard particulate biological contaminants and often filter them from the percolating leachate. The chemical contaminants, however, may be carried by the leachate water to the water table where they enter the ground-water flow system and move according to the hydraulics of that system. Thus, the potential for pollution in the hydrologic system depends upon the mobility of the contaminant, its accessibility to the ground-water reservoir, and the hydraulic characteristics of that reservoir.

The character and strength of the leachate are dependent in part upon the length of time that infiltrated water is in contact with the refuse and the amount of infiltrated water. Thus, in areas of high rainfall the pollution potential is greater than in less humid areas. In semiarid areas there may be little or no pollution potential because all infiltrated water is either absorbed by the refuse or is held as soil moisture and is ultimately evaporated. In areas of shallow water table, where refuse is in constant contact with the ground water, leaching is a continual process producing maximum potential for ground-water pollution.

The ability of the leachate to seep from the refuse to the ground-water reservoir is another factor in the degree of pollution of an aquifer. Permeable soils permit rapid movement; although some filtering of biological contamination may take place, the chemical contamination is generally free to move rapidly under the influence of gravity to the water table. Less permeable soils, such as clays, retard the movement of the leachate, and often restrict the leachate to the local vicinity of the refuse. Under such conditions, pollution is frequently limited to the local shallow ground-water reservoir and contamination of deeper lying aquifers is negligible.

Leachate that does reach the water table and enters an aquifer is then subject to the hydraulic characteristics of the aquifer. Because the configuration of the water table generally reflects the configuration of the land surface, the leachate flows downgradient under the influence of gravity from upland areas to stream valleys, where it discharges as base flow to the stream systems. The rate of flow is dependent upon the permeability of the rock material of the aquifer and on the slope of the water table. In flat areas or areas of gentle relief, minor local topographic variations may have no effect on the configuration of the water table, and movement of ground water may be uniform over large areas.

In some places dipping confined aquifers crop out in upland areas and thus are exposed to recharge. Contaminants entering the aquifer in these areas move downgradient into the confined parts of the aquifer. Although there is usually some minor leakage to confining beds above and below the aquifer, the contaminants in general will be confined to the particular aquifer, and water-supply wells tapping that aquifer will thus be subject to contamination to the extent that the contaminants are able to move from the outcrop to the wells.

Optimum conditions for pollution of the ground-water reservoir exist where the water table is at or near land surface, subjecting the solid waste to continual direct contact with the water. Such conditions commonly exist where abandoned quarries that penetrate the ground-water reservoir are used as refuse-disposal sites. The continual contact of the water with the refuse produces a strong leachate highly contaminated both biologically and chemically. Under hydrogeologic conditions of permeable materials and steep hydraulic gradients, the leachate may move rapidly through the ground-water system and pollute extensive areas. The hydrologic effects of solid-waste disposal in four geologic environments are shown in Figure 2.

Figure 2A illustrates a waste-disposal site in a permeable environment. The waste is shown in contact with the ground water in a permeable sand-and-gravel aquifer underlain by confining beds of relatively impermeable shale. In this case, the potential for pollution is high because conditions of both high infiltration and direct contact between wastes and ground water exist. Because of the permeability of the aquifer, the contaminants move downgradient with the water in the aquifer and are diffused and diluted during this movement. In areas where the water table is below the bottom of the waste material, the degree of contamination is lessened because the wastes are no longer in direct contact with the ground water. In this case, leachate from the wastes moves vertically through

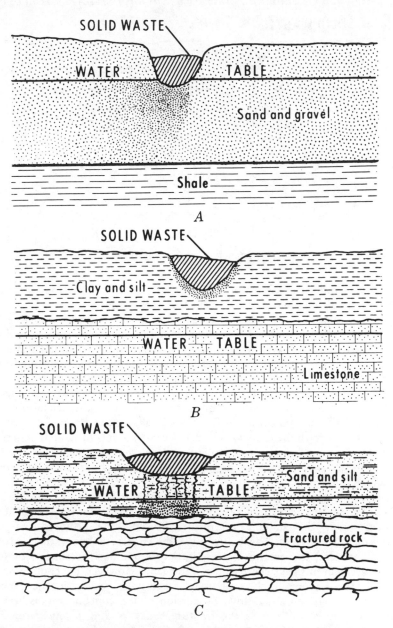

FIGURE 2. Effects on ground-water resources of solid-waste disposal at a site (A) in a permeable environment, (B) in a relatively impermeable environment, (C) underlain by a fractured-rock aquifer, and (D) underlain by a dipping-rock aquifer.

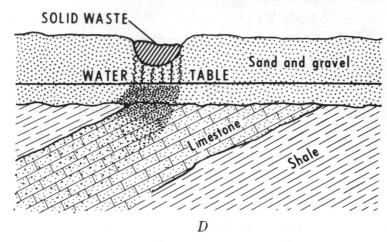

D

FIGURE 2. Continued

the zone of aeration to the water table. It then enters the ground water and moves downgradient as in the case of a shallow water table.

Figure 2B illustrates a waste-disposal site in a relatively impermeable environment. In humid areas, the water table may be near land surface, and the disposed waste may or may not be in direct contact with the ground water. In the illustration, ground water is shown confined to the underlying limestone aquifer. The relative impermeability of the overburden prevents significant infiltration of the rainfall; consequently there • is only minor leaching of contaminants from the wastes. Pollution is confined locally to the vicinity of the waste-disposal site; movement in all directions is inhibited by the inability of the water to move through the tight soils. If significant amounts of rainfall penetrate the wastes, a local perched water table may develop in the vicinity of the fill, and that water will likely be highly contaminated, both chemically and bacteriologically.

Figure 2C illustrates a waste-disposal site above a fractured-rock aquifer. The position of the water table in the overburden relative to the waste-disposal site is dependent upon the amount of infiltration and the geometry of the ground-water flow system. The water table shown here is below the body of waste. In this case, the potential for pollution is not high because of limited vertical movement of the leachate to the water table. However, the contaminants that reach the fractured-rock zone may move more readily in the general direction of the ground-water flow. Dispersion of the contaminants is limited because the flow is confined to the fracture zones. A thin, highly permeable overburden with a shallow water table (similar to that shown in Figure 2A) overlying the fractured rock would provide an ideal condition for wide spread ground-water pollution.

Figure 2D illustrates a waste-disposal site in a geologic setting in which dipping aquifers are overlain by permeable sands and gravels. In this illustration, the waste-disposal site is shown directly above a permeable limestone aquifer. Here leachate from the landfill travels through the sand and enters the limestone aquifer as recharge. Again, the strength of the leachate depends in part upon whether the water table is in direct contact with the waste. Leachate will move downgradient with the ground-water flow in both the sands and gravels and the limestone, as shown in the illustration. If the waste-disposal site were located above the less permeable shale, most of the leachate would move downgradient through the sand and gravel, with very little penetrating the relatively impermeable shale as recharge. However, in its downgradient movement, it would enter any other permeable formations as recharge.

A high pollution potential exists also where waste-disposal sites are located on flood plains adjacent to streams. Water-table levels generally are near land surface in flood-plain areas, especially during the usual period of high water in winter and spring throughout much of the humid areas of the United States. In such environments the water may have contact with the refuse for extended periods, giving rise to concentrated leachate. The contaminated water moves through the flood-plain deposits and discharges into the stream during low-flow periods when the bulk of the streamflow is from ground-water discharge. The degree of pollution of the stream depends upon the concentration of the leachate, the amount of leachate entering the stream, and the available streamflow for dilution.

Hydrologic Considerations in Site Selection

It is obvious that our current national policy of pollution abatement and protection of our natural environment requires full consideration of the water resources in selection of sites for solid-waste disposal. To date, with few exceptions, these considerations have been on a local scale, dealing primarily with the hydrological characteristics of the immediate site.

The American Society of Civil Engineers in a manual on sanitary landfill (American Society of Civil Engineers, 1959) discussed site selection from a hydrologic standpoint as follows:

> In choosing a site for the location of a sanitary landfill, consideration must be given to underground and surface water supplies. The danger of polluting water supplies should not be overlooked.

The report states further that:

> Sufficient surface drainage should be provided to assure minimum runoff to and into the fill. Also, surface drainage should prevent quantities of water from causing erosion or washing of the fill. . . . Although some apprehension has been expressed about the underground water supply pollution of sanitary landfills, there has been little, if any, experience to indicate that a properly located sanitary landfill will give rise to underground pollution problems. It is axiomatic, of course, that when a waste material is disposed of on land, the proximity of water supplies, both underground and surface, should be considered. . . . Also, special attention should be given areas having rock strata near the surface of the ground.

For example, limestone strata may have solution channels or crevices through which pollution contamination may travel. Sanitary landfills should not be located on rock strata without studing the hazards involved. In any case, refuse must not be placed in mines or similar places where resulting seepage or leachate may be carried to water-bearing strata or wells. . . . In summary, under certain geological conditions, there is a real potential danger of chemical and bacteriological pollution of ground water by sanitary landfills. Therefore, it is necessary that competent engineering advice be sought in determining the location of a sanitary landfill.

Consideration of hydrology in site selection is required by law in several States. Section 19-13-B24a of the Connecticut Public Health Code requires that:

No refuse shall be deposited in such manner that refuse or leaching from it shall cause or contribute to pollution or contamination of any ground òr surface water on neighboring properties. No refuse shall be deposited within 50 feet of the high water mark of a watercourse or on land where it may be carried into an adjacent watercourse by surface or storm water except in accordance with Section 25-24 of the General Statutes which require approval of the Water Resources Commission.

The rules and regulations of the Illinois Department of Public Health require that:

The surface contour of the area shall be such that surface runoff will not flow into or through the operational or completed fill area. Grading, diking, terracing, diversion ditches, or tilling may be approved when practical. Areas having high ground water tables may be restricted to landfill operations which will maintain a safe vertical distance between deposited refuse and the maximum water table elevation. Any operation which proposes to deposit refuse within or near the maximum water table elevation shall include corrective or preventive measures which will prevent contamination of the ground-water stratum. Monitoring facilities may be required.

Other States have similar regulations.

A common denominator in these sets of recommendations is the general concern for the onsite pollutional aspect. This is characteristic of most current approaches, especially from the engineering and legislative viewpoints. Another characteristic is the restrictive approach to the problem. Hydrologic conditions are documented under which disposal of solid wastes is either discouraged or prohibited. In general, the pollutional problem is treated more in local than in regional context. These are, of course, important considerations and should be followed in any site selection. In fact, even stronger guidelines are desirable to the extent of requiring detailed knowledge of the extent and movement of potential pollution at any site before the site is activated.

The water resource, however, must be considered also as a regional resource, not just a localized factor. As such, it should be considered in a regional concept in its relation to solid-waste disposal. This, of course, requires that adequate regional information on the water resource is available. Given such information, the planner can weigh all available alternatives and insure that the final site selection is compatible with comprehensive regional planning goals and environmental protection. The Northwestern Illinois Planning Commission followed this comprehensive approach in developing its recommendations on refuse-disposal needs and practices in northeastern Illinois (Sheaffer and others, 1963).

It is, of course, quite possible that, in the comprehensive approach, some otherwise optimum sites for solid-waste disposal may be only marginally acceptable from a hydrologic viewpoint. Under such conditions detailed information on the hydrology should be obtained and detailed evaluations made of the impact of the potential waste disposal before the site is put into use; the actual impact should then be monitored during and after use. In general, although such studies are desirable for all solid-waste-disposal sites, they are essential where geologic, hydrologic, or other data indicate a possibility of undesirable pollutional effects.

The problem of solid-waste disposal is one of the most serious problems of urban areas. The ever-increasing emphasis on protection and preservation of natural resources through regional planning is evident today. The implementation of these commitments and goals can insure adequate protection of vital water resources from pollution by disposal of solid wastes.

References

1. American Public Works Association. 1966. Municipal refuse disposal. Chicago. Public Adm. Service, 528 p.
2. American Society of Civil Engineers. Committee on Sanitary Landfill Practices. 1959. Sanitary landfill. Am. Soc. Civil Engineers Eng. Practices Manual 39, 61 p.
3. Anderson, J. R., and J. N. Dornbush. 1967. Influence of sanitary landfill on ground water quality. J. Am. Water Works Assoc., April 1967, p. 457-470.
4. California Water Pollution Control Board. 1961. Effects of refuse dumps on ground water quality. California Water Pollution Control Board Pub. 24, 107 p.
5. Hughes, G. M. 1967. Selection of refuse disposal sites in northeastern Illinois. Illinois Geol. Soc. Environmental Geology Note 17, 18 p.
6. Hughes, G. M., R. A. Landon, and R. N. Farvolden. 1969. Hydrogeology of solid waste disposal sites in northeastern Illinois. U.S. Department of Health, Education, and Welfare, Public Health Service, 137 p.
7. Sheaffer, J. R., B. von Boehm, and J. E. Hackett. 1963. Refuse disposal needs and practices in northeastern Illinois. Chicago. Northeastern Illinois Metropolitan Area Planning Comm. Tech. Rept. 3, 72 p.

Part Four

Examples of Ground-Water Pollution

E xamples of ground-water pollution have appeared in the literature for scores of years. The reports included in this section are not unique but are intended to show typical problems, their cause, effect and, in some instances, methods of investigation.

In nearly every situation it is apparent that contamination occurred through ignorance and neglect and not by intent. Wastes, intentionally or unintentionally spread over the land surface, dumped into pits, or injected into the ground can cause significant problems. Because ground water moves very slowly, the waste may remain undetected for decades. Moreover, once detected, the situation may require many more years to be corrected.

Water may be contaminated with biological matter, chemical wastes, or a combination of both, and consumption of this water may lead to significant health problems, either acute or chronic. It seems strange that many people tend to ignore the contamination of water supplies, particularly ground water, where human wastes are involved. The relationships between human wastes, pathogenic organisms and disease have been known for decades. Unquestionably, health problems exist, not only in rural areas, but more particularly in suburbia to which the rapidly expanding population is retreating. In such areas community water and sewage systems often lie several years in the future. Meanwhile, septic

Photo opposite courtesy of *Minneapolis Tribune*, Minneapolis, Minn.

147

tanks or cesspools can contaminate the ground water over large areas. Fortunately, as pointed out by J. C. Romero, bacteria and viruses tend to move only relatively short distances in a porous media and this, no doubt, is why there are so few epidemics directly traceable to water supplies from properly constructed wells. Unlike bacteria and viruses, soluble substances such as phenols, chromium, nitrate, and oil-field brines may move great distances in the ground; they may be diluted but are commonly not absorbed by the porous media. Instances such as these are described by M. Deutsch, G. Walton, R. H. Bogan, J. E. Vogt, and W. A. Pettyjohn.

Alternate waste-disposal techniques may lead to other problems. Graham Walton's report briefly mentions that waste pits at the Rocky Mountain Arsenal were to be replaced with a deep disposal well. Following completion of the well and the injection of large quantities of liquid waste into it, earthquakes began to occur in greater frequency in the Denver area; they have been attributed to the injection well.

Incidents of Chromium Contamination of Ground Water in Michigan

Morris Deutsch*

Our ground water resources in some areas have been contaminated in many different ways and by many different contaminants. A review of numerous incidents of ground water contamination in Michigan by the author (1) revealed how easily aquifers can become contaminated, how widespread and costly such contamination already is, and how difficult it is to remove contamination once introduced. The review also outlined principles controlling the entry and movement of wastes in aquifers. Among the most serious (and interesting) cases reviewed were those involving entry of chromium compounds, especially hexavalent chromium, into aquifers used as sources of public supply.

The chromium-contamination incidents demonstrate a few of the ways wastes may enter an aquifer. These incidents serve as examples of the seriousness of ground water contamination. They provide us with an opportunity to examine the hydrogeologic controls that govern the underground movement of the contaminants and also give us some insight concerning the extent and duration of the effects of contamination of our ground water resources.

Electroplating, especially chrome plating, is a relatively small but important industry in Michigan. The industry has the problem of disposing of plating wastes, which are variable in character and usually include hexavalent chromium, cyanide, and caustic soda in the rinse waters. Of particular concern with respect to plating wastes is the fact that minute concentrations of hexavalent chromium and cyanide render water unfit for human consumption. According to the U.S. Public Health Service (2), "hexavalent chromium in excess of 0.05 ppm (1 part to 20 million) shall constitute grounds for rejection of the (public-water) supply." The toxicology laboratory of the Michigan Department of Health reported that chromium in the amount of 1 ppm may have a detrimental effect on the human nervous system and kidney tissues and that chronic illnesses may result. The chromate imparts a yellow tinge to the water in which it is dissolved. Cyanide in any amount whatsoever is intolerable in water used for public supply.

In the past, disposal of electroplating wastes to streams has created serious

Reprinted from *Public Health Service Technical Report W 61-5, 1961, pp. 98-104.*
*Water Resources Division, U.S. Geological Survey, Washington, D.C.

hazards to the public health. As an alternative to surface disposal of electro-plating wastes, some concerns have attempted to dispose of the wastes in infiltration pits. This practice has some merit in that the hazard from cyanide reportedly is largely eliminated. The Michigan Water Resources Commission observed that although they "have encountered a number of ground-water-pollution problems involving electroplating wastes, no instance has occurred . . . where cyanide could be traced in wells any distance from the point of disposal. This is accounted for by the various reactions to which cyanide is subjected by subsurface formations."

Percolation from Ponds

Disposal of the waste to the ground has not solved the chromium-con-tamination problem, however, since chromium is not completely removed from the water by the rock materials through which it percolates. Almost all the suspended solid material is filtered out by the first few inches of soil, but the water containing dissolved chromium moves through the permeable materials and reaches the aquifers. In general, the movement is downward, although some water is dispersed laterally by capillarity or deflected by lenses or zones of low permeability. Eventually, the contaminated water enters the upper part of the underlying aquifer (Figure 1). (All the figures included in this paper are schematic and are not based on field data.) The liquid waste tends to form a mound on the water surface and moves radially from the mound. The direction and velocity of underflow of the waste is controlled principally by the gradient and the permeability of the aquifer materials. Once the chromium is introduced, the aquifer is unfit as a source of potable water for a prolonged period of time because of the generally slow movement of ground water. It is not known whether natural flushing action or dewatering by pumping will in time remove the chromium from the aquifer.

Douglas Incident

Several instances of chromium contamination have occurred in the State. In 1947 the Michigan Department of Health (3) was notified that water from wells tapping the glacial drift in the western part of the village of Douglas in Allegan County had turned yellow. The wells were removed from service, pending analysis of a water sample. The analysis revealed a chromate content of 10.8 ppm or more than 200 times the concentration of hexavalent chromium recommended by the Public Health Service as the maximum safe limit in public supply systems.

The source of contamination was quickly located. About 3 years before the contamination appeared, a metal-plating concern began discharging chrome-plating wastes into an infiltration pit and the surrounding overflow area about 1000 feet south of the western wells and 2500 feet southwest of the eastern well at Douglas. Discharge of the plating wastes had resulted in contamination of the glacial-drift aquifer for at least 1000 feet in one direction from the pit and to a depth of at least 37 feet. It had taken about 3 years for the waste to migrate

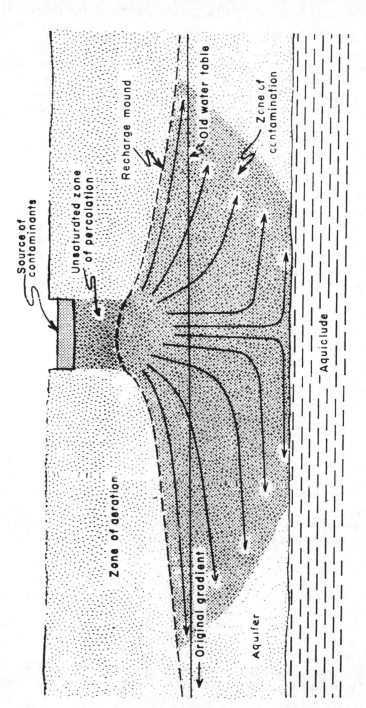

FIGURE 1. Schematic diagram showing percolation of contaminants through zone of aeration and into isotropic aquifers.

1000 feet at a rate of about 1 foot per day along the gradient created by pumping of the western wells. Health Department personnel estimated at the time that if disposal to the pit were stopped immediately it would be at least 6 years before the aquifer in the vicinity of the west wells would be free of chromate. Although the 1947 analyses of water from the eastern well showed no chromate content, water from the well was analyzed periodically as a safeguard. However, the wells of a local dairy were found to be contaminated. The Health Department requested that the Village Council condemn all private wells in the village, since there was no practical way of observing the quality of the water from each well.

Bronson Incident

Since 1939 electroplating industries at Bronson, in southern Michigan, have experienced difficulty in disposing of electroplating wastes (4). Originally, the wastes were discharged into the city's sewer system, which emptied into a county drain and a creek. Contamination of these water-courses resulted in the death of fish and cattle below Bronson as a result of ingestion of cyanide. The city subsequently issued an ordinance to prohibit discharge of toxic wastes to the city's sewer systems. All the plating wastes of the principal company involved subsequently were discharged to two ponds. In 1942 it was found that the dikes around the ponds were unsafe, and inspection of the flow from the sewer system revealed a faint yellow color, characteristic of chromium-waste contamination. The chromium probably resulted from leakage of water from the ponds, both above and below ground, or use of the sewer system for waste disposal. Subsequent cases of surface water contamination were reported.

In 1949 ground water contamination at Bronson was revealed when the owner of a domestic well observed a "greenish tinge" in his well water. The well was 75 feet from the sewer that carried plating wastes to the disposal ponds. In December of 1949 the Water Resources Commission collected samples of water from one of the ponds, from the domestic well, and from a well at the County Highway Garage between the pond and the domestic well. The domestic well was 14 feet deep, and the water level was 8 feet 6 inches below the land surface. The well at the garage was 33 feet deep, and the water level was 8 feet 2 inches below the land surface. Both wells tapped the same shallow glacial-drift aquifer. Results of the analyses made by the Michigan Department of Health are given in Table 1.

It was evident that the part of the aquifer tapped by the domestic well was contaminated at the time of sampling but that the deeper part, which was tapped by the Highway Garage well, was not. A check of the sewer lines revealed that they were in good condition and were not contributing contamination to the aquifer. Evidently, the plating wastes were moving directly from the disposal ponds. By December 29, 1949, the chromium content of the domestic well had risen to 3.5 ppm.

Several interesting hydrologic observations can be made, based on the instance of chromium contamination at Bronson. The chemical analyses revealed

TABLE 1. Results of analyses of water from electroplating
plant disposal pond and two nearby wells.

	Pond	Garage well	Domestic well
Cyanide, ppm	15.6	0	Trace
Chromium, ppm	6.0	0	2.0
Nickel, ppm	49.0	0	Trace
Copper, ppm	12.0	0	0
pH	6.65	7.5	7.5

that the 33-foot well at the County Garage was not contaminated at the time the sample was collected, although the well is between the contaminated shallow domestic well and the disposal pond. This shows that the chromium contamination was not uniformly distributed throughout the aquifer. The contaminant may have been confined to the upper part of the ground water body and only slightly dispersed in traveling through the aquifer. Movement to the deeper well could have been impeded in part by lenses of low permeability within the aquifer.

The natural gradient of the water table in the shallow-drift aquifer was reported to be northwestward, but the contaminated well was southwest of the disposal pond. This indicates that a ground water mound was built up that moved the wastes in a direction opposite to the natural gradient. Underground movement of contaminated water would tend to be radial from beneath the pond (Figure 2). The degree of buildup of the mound and hence the distance the water moves opposite to the natural gradient is controlled by the slope of the natural gradient, the physical character of the aquifer, and the quantity of contaminated water introduced. (Pumping of wells also would result in local gradients opposite to the natural gradient. Thus, areas upgradient from sources of contamination are not necessarily protected from pollution.)

After more than a decade, there is concern that new wells proposed to be drilled to a deeper aquifer by the city also might be subject to chromium contamination. The lower aquifer is reported to be separated from the contaminated upper aquifer by a "thick impervious clay stratum." The concern that deeper wells may be contaminated may be warranted, however. Aquifer tests made in numerous areas of the State by the Federal and State Geological Surveys seldom reveal artesian conditions perfect enough to completely shut off all interaquifer movement of water. Study of the well logs or even visual inspection of a clay layer is inadequate to determine the impermeability of clayey materials. Laboratory analysis of permeability or extensive aquifer testing would be necessary to determine the hydraulic characteristics of the "impervious clay stratum." Further, it would have to be determined whether the confining layer is penetrated by wells with rusted or broken casings that would permit leakage of contaminants to the lower aquifer (Figure 3).

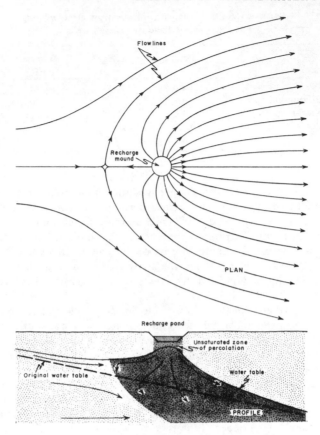

FIGURE 2. Diagrams showing lines of flow
from mound on sloping water table.

Leaching from the Land Surface

Kent County Incident

Several incidents of ground water contamination have resulted from uncontrolled spilling or spreading of chromium-bearing substances on the land surface. Disposal of chrome-bearing liquids or soluble solids that can percolate to an underlying aquifer are a potential hazard, as illustrated by Figure 4. An incident in Kent County (5) indicates how chromium contaminants can be spread on the surface through "normal" activities. During the winter of 1955-56, some of the snow along a roadside in the area turned yellow. Investigation by a township engineer revealed that the Kent County Road Commission was using salt to melt ice and snow on the county roads. The salt was treated with a chromium-base rust inhibitor to allay county residents' complaints concerning

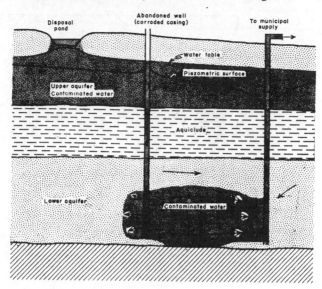

FIGURE 3. Generalized diagram showing interformational
leakage by vertical movement of water through wells.

rapid corrosion of automobile bodies. Samples of snow had a chromium content
of 10 ppm. Fortunately, the township engineer recognized the potential hazard
to the water supplies of the area and notified the County Road Commission; the
use of chromium-treated salt was promptly discontinued.

A unique case of ground water contamination by chromium occurred in the
city of Grandville west of Grand Rapids (6). In this incident the city drilled a
public supply well in the glacial-drift deposits along the Grand River. To protect
the well from flooding during periods of high water, the casing was extended
several feet into the air and the land surface was raised by filling with sand and
gravel. In time, chromium was detected in the water. This resulted in considerable
consternation, since there was no apparent source of chromium contamination
in the vicinity. Investigation by the Grandville Superintendent of Water revealed
that the sand and gravel fill used to raise the land surface at the well was taken
from a former dumping grounds for electroplating wastes! The river was in flood
stage shortly before the contamination appeared. Water from the river obviously
had leached the chromium from the fill and carried it into the aquifer.

Discharge of Chrome-Laden Dust

Wyoming Township Incident

For several years Wyoming Township in Kent County had difficulty obtaining
from one of its well fields water that was free from chromium. An electroplating
firm on adjacent property was the apparent source of the chromium

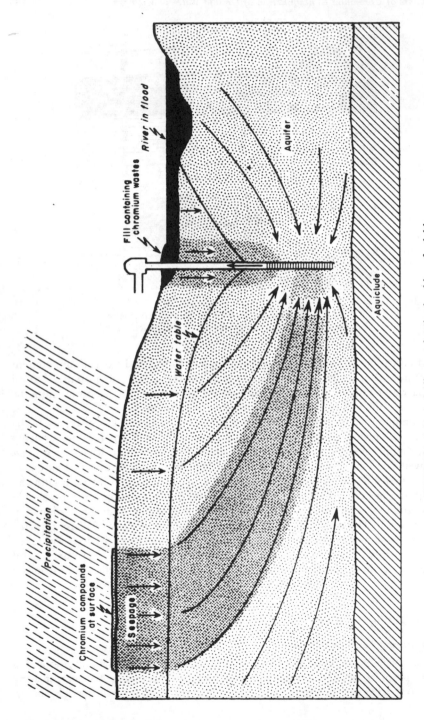

FIGURE 4. Schematic diagram showing leaching of soluble
contaminants into aquifer from precipitation and flood waters.

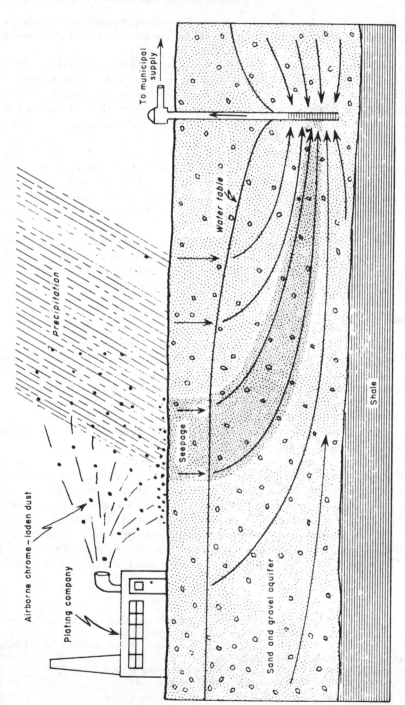

FIGURE 5. Diagram showing possible mode of entry of airborne wastes to aquifer.

To municipal supply

Airborne chrome – laden dust

Plating company

Precipitation

Water table

Seepage

Sand and gravel aquifer

Shale

contamination. Accordingly, the firm retained a consulting engineer to study the problem and report on the possibility of further contamination and of means of abatement (7).

The engineer concluded in his study that airborne chromium could have been introduced to the aquifer in several ways. Chrome-laden dust was discharged through ventilators on the roof. Some of the dust settled to the ground, where it accumulated until rainfall washed it down to the water table, as depicted in Figure 5. A general relationship was shown between precipitation and chromium contamination in the township wells.

A dry well at the site, into which water from the roof on the plant drained, was another likely source of intermittent contamination. Some of the dust probably was washed out of the air and onto the roof by precipitation. The dust may have flowed down a rainspout into the dry well and then infiltrated to the aquifer, through which it subsequently migrated to the well field in response to pumping.

In addition, the Michigan Geological Survey reported that chrome-powder residue was present in or on containers left in the yard. Some of this hazardous powder may have been spilled on the ground from where it readily could have been leached and carried into the underlying aquifer by subsequent rainfall.

Summary

Several incidents of chromium contamination of ground water have occurred in Michigan. The most serious of these resulted from disposal of electroplating wastes to ponds or settling basins. Other incidents of ground water contamination were caused by spreading of chromium-treated salt to melt snow, by use of chromium-contaminated land fill, and possibly even by settling of chromium-laden dust from the air. Fortunately, chromium contamination of ground water has been of small areal extent, and no human fatalities or serious illnesses are known to have occurred.

The modes of entry of chromium-bearing wastes to aquifers are well recognized by Michigan public agencies concerned with water resources and public health. Through their efforts, contamination by electroplating industries virtually has been eliminated. In addition, very few incidents of contamination by chromium from other sources have been reported. Because minute concentrations of hexavalent chromium in water are highly toxic, aquifers must be constantly protected from future contamination. To help protect our ground water resources, the public generally must be made aware of the hazards involved in the disposal of toxic wastes.

References

1. Deutsch, Morris, 1960. Ground-water contamination and legal controls in Michigan. U.S. Department of the Interior, Geological Survey open-file rept.
2. U.S. Department of Health, Education and Welfare, Public Health Service. 1946 drinking water standards. Public Health Service Repts. 6 (11):371-384.
3. Michigan Department of Health. 1947. Mich. Water Works News, v. 7, no. 3, 1947.
4. L. A. Darling v. Water Resources Commission (341 Mich. 654).

5. Michigan Department of Health. 1958. Inhibitors used in snow removal. Mich. Water Works News, v. 23, no. 2, 1958.
6. Michigan Department of Health. 1956. Unique pollution of a well by chromium. Mich. Water Works News, v. 21, no. 3, 1956.
7. Report to the Grand Rapids Brass Company on industrial wastes and water supply. 1956. Williams and Works mimeo. rept.

Public Health Aspects of the Contamination of Ground Water in the Vicinity of Derby, Colorado

Graham Walton*

The Rocky Mountain Arsenal is located just northeast of Denver, Colorado (Figure 1). Started in 1943, the arsenal was operated by the Chemical Corps for several years for the production of chemical warfare agents. More recently, the industrial facilities have been leased to the Shell Oil Company, which has utilized them to manufacture insecticides.

From 1943 through September 1955 wastes from various chemical processes were discharged to Reservoir A (the locations of the reservoirs are shown in Figures 1 and 2). During part of that period wastes from the chlorine-processing plant were discharged to Reservoir C. Since early October 1955, all industrial wastes have been discharged into the 96-acre, asphalt-membrane-lined, evaporation Reservoir F.

Wastes from the unlined holding ponds have seeped into the ground and contaminated the shallow ground water throughout approximately 5 square miles of the South Platte River valley immediately northwest of the arsenal property. This is an area of farms and some suburban housing. The region is semi-arid, and water for crop irrigation has been obtained largely from shallow wells. A second source of water is a deep artesian aquifer, which yields only limited quantities. Most of the domestic wells tap the latter source, since the water from the shallow aquifer has always been highly mineralized.

History of Contamination

The first indication that the ground water had become contaminated was damage to crops irrigated with water from shallow wells. Such crop damage was observed at the Newson farm in 1951, at the Powers and Munson farms in 1952, and at the Yamamoto and Miller farms in 1953.

Complaints and subsequent claims for damages led to the Chemical Corps' engaging a firm of consulting engineers to investigate the problem. Their report

Reprinted from *Public Health Service Technical Report* W61-5, 1961, pp. 121-125.
*When this report was written Graham Walton was in the Research Branch, Division of Water Supply and Pollution Control, U.S. Public Health Service in Cincinnati, Ohio. He is now retired.

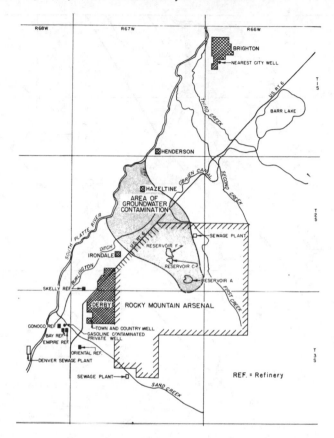

FIGURE 1. Derby, Colorado, area investigated in connection
with contamination of shallow ground water.

recommended and resulted in a substantial reduction in the volume of wastes, and, starting in October 1955, the retention of all wastes in Reservoir F. Another result was a contract with the University of Colorado to undertake plant bioassay, chemical, and geological studies to determine the identity and source of any contaminants causing crop damage. These and other studies emphasized the effect of the contaminated water on agricultural uses. It was not until 1959, when the State of Colorado requested that the Public Health Service make a reconnaissance survey, that serious concern developed over the use of such waters for domestic purposes.

Public Health Study

At the time of the Public Health Service survey, contaminants known to have been present in certain shallow well waters taken from within the area included

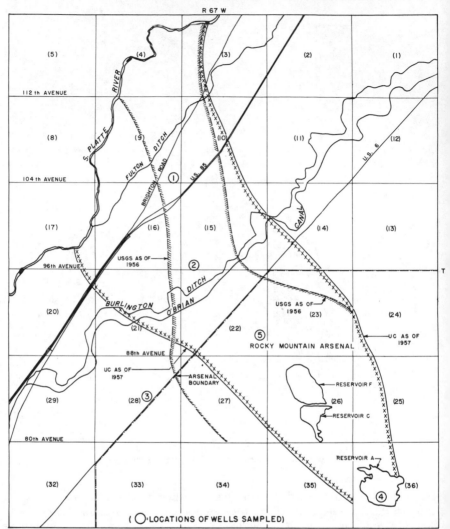

FIGURE 2. South Platte River basin, Colorado—suspected areas of
shallow ground water contamination shown as of 1956 and 1957.

chlorides and chlorates. The weedicide 2,4-D was known to have been isolated
from wastes in the holding pond, and plant bioassay studies had indicated that it
or other phytotoxic organic substances had been present in some of the shallow
well waters. Still other contaminants known to have been in wastes discharged
from the arsenal's operations were salts of phosphonic acid, fluorides, and
arsenic.

During August 1959, visits to 50 of some 150 homes in the area provided data on 23 domestic water supplies. Eighteen were deep well supplies, which were reported satisfactory wherever a member of the household could be interviewed. Five dwellings were served by shallow wells. Waters from such wells were used for drinking and for culinary purposes at three residences, at two of which the water was reported to have bad taste and odor.

Previous studies by the U.S. Geological Survey (1) and the University of Colorado (2) had established the general area of shallow groundwater contamination. The isochloride line for the 200-mg/l concentration as of June 1956 is shown in Figure 2. Since shallow ground waters from this area normally contain more than 100 mg/l of chloride, the 200-mg/l concentration is evidence that at least 3 to 5 percent of the water within the area originated from the highly contaminated ground water underlying the arsenal property. Also shown in Figure 2 are boundaries of the area throughout which the shallow well waters had exhibited phytotoxic characteristics, as determined by the University of Colorado studies through 1957.

Among the recommendations made in the report on this study (3) were that steps be taken immediately to determine present boundaries of the area in which the shallow ground water had become contaminated, to analyze all domestic water supplies from wells within that area, and to provide written notice to the owners of contaminated wells that the water is "unsafe for drinking or culinary uses." Chloride concentrations in excess of 200 mg/l were to be considered as evidence of contamination until more adequate information could be developed concerning the contaminants present and their concentrations. Other recommendations provided for a monitoring program, toxicological studies to better establish safe permissible limits of certain contaminants in potable water, and an investigation of the sludge accumulation in Reservoir A.

Subsequent activities of the Public Health Service have been limited to analyses of five well water samples collected between October 29 and November 3, 1959 (4), a study of the toxicity of chlorates, and consultation services to the Colorado State Department of Health.

Table 1 summarizes the results of the analyses of the well waters. Figure 2 shows that Well No. 3 is adjacent to, but outside, the area influenced by seepage from the ponds receiving wastes from operations at the Rocky Mountain Arsenal. All other wells are within the area contaminated. Although such contaminants as chlorates, phosphonates, and 2, 4-D were, if present, well within the tentative permissible limits recommended for drinking water, other substances—solids, hardness, alkalinity, fluorides, and chlorides—showed that these waters have abnormal characteristics that could only be associated with wastes from the operations at the Arsenal. Exploratory spectrographic analyses also showed the presence of abnormal concentrations of iron, manganese, and molybdenum, which indicated the corrosive characteristics of these waters.

Waters from the four wells were considered unsuitable for domestic water supplies on the basis of the analytical results for the samples submitted. Further consideration of the source of the contamination, the possible nature of the contaminants that might be present, and the history of damage to crops when

TABLE 1. Comparison of analytical results for waters from wells in or adjacent to contaminated area near Derby, Colorado, with 1946 PHS drinking water standards. (All results except pH in milligrams per liter)

		Well number and sampling date				
Analysis	1946 PHS Standards	No. 3, in uncontaminated area (11/2/59)	No. 4, in bed of reservoir (11/3/59)	No. 5, on Rocky Mtn. Arsenal (11/3/59)	No. 2, irrigation well (10/29/59)	No. 1, in Hazeltine area (10/29/59)
Solids						
Total	1,000[a]	492	10,000	2,800	3,760	2,620
Volatile		86	965	446	980	808
Hardness (total as $CaCO_3$)		212	1,990	280	924	928
Calcium	125[b]	85	369	77	369	122
Magnesium		0	258	21	0	152
Alkalinity						
Total (as $CaCO_3$)		227	700	827	372	262
Hydroxide (as $CaCO_3$)		0	13	23	0	0
Sulfate (as SO_4)	250[b]	104	3,000	249	435	455
Fluoride	1.5[b]	0.6	4.0	2.8	1.2	1.2
Chloride	250[b]	40	1,410	864	1,320	764
Sodium		51	2,200	880	580	300
Potassium		3.4	12	7.9	6.9	5.0
Phosphate						
Ortho (as PO_4)		0.12	4.5	3.4	0.08	0.16
Poly (as PO_4)		0.04	0.24	0.16	0.22	0.34
Phosphonate (as PO_4)		0.0	0.0	0.1	2.1	0.0
Chlorates (as ClO_3)		< 1	< 1	< 1	< 1	< 1
2,4-D		< 0.2	< 0.2	< 0.2	< 0.2	< 0.2
ABS		0.1	0.7	0.6	0.6	0.4
pH		8.3	8.4	8.6	8.0	8.0

[a]Recommended maximum limit 500 mg/1, permitted 1,000 mg/1.

such water had been used for irrigation required that these waters be rejected as domestic water supplies.

Subsequent Developments

Since 1959, the Chemical Corps has hauled water for domestic use of those householders formerly using contaminated well water. It is understood also that provision has been made to facilitate processing of limited claims for damages, such as the expense incurred in drilling a well into the deeper, uncontaminated aquifer.

Both the State Department of Health and the Chemical Corps are monitoring selected wells. Available information reveals no increase in the area with contaminated shallow ground water.

A survey of the industrial waste system has been made and a contract awarded for the construction of a plant to pretreat the wastes prior to their injection into a deep well. All wastes will be discharged to Reservoir F, where chemical nutrients will be added to induce "a complete biological cycle." Subsequent treatment will consist of flocculation, settling, pressure filtration, and disinfection. The clarified liquid will be deoxygenated before injection into the ground.

A contract was let February 2, 1961, for construction of a deep injection well, the plans for which have been approved by the Colorado State Department of Health. This well will be tripled-cased throughout the upper strata and double-cased to a depth of 2,000 feet. The 8-3/8-inch-OD inner casing will extend to a depth of approximately 10,000 feet. Each casing will be cemented in place. Wastes will be injected through a 5-1/2-inch tube extending to the stratum in which they will be discharged. The annular opening between this tube and the casing will be filled with clean water. Since the injection pressure at the well head will be high — up to 2,000 psi, leakage of waste into the annular opening will be detectable by an increase in the reading of a recording pressure gage installed for that purpose.

Consideration also is being given to pumping ground water from wells in the more contaminated areas to waste into the South Platte River. This would be done only at times when the flow of the river provides sufficient dilution to maintain the concentrations of the contaminants within permissible limits.

References

1. Petri, L. R. and R. O. Smith. 1956. Investigation of the quality of ground water in the vicinity of Derby, Colorado. U.S. Department of Interior, Geol. Survey, Water Quality Division.
2. Bonde, E. K. 1958. Research on phytotoxic materials, Contract DA-05-021-401-CHL 10,092. Department of Biology, University of Colorado.
3. Walton, Graham. 1959. Public health aspects of the contamination of ground water in the South Platte River basin in the vicinity of Henderson, Colorado, August, 1959. U.S. Department of Health, Education, and Welfare, Public Health Service, Robert A. Taft Sanitary Engineering Center.
4. Walton, Graham, 1960. Report on analyses of water samples from Rocky Mountain Arsenal area, Denver, Colorado. U.S. Department of Health, Education, and Welfare, Public Health Service, Robert A. Taft Sanitary Engineering Center.

Water Pollution by Oil-field Brines and Related Industrial Wastes in Ohio

Wayne A. Pettyjohn*

Introduction

Since drilling of the first oil well at Titusville, Pennsylvania, in 1859, pollution problems related to the disposal of oil-field brines have confronted both the petroleum industry and the general public as well. Practically every oil-producing state has enacted laws that regulate the drilling and plugging of wells and the disposal of brines. Many of these laws are the direct result of ground-water or surface-water contamination.

The serious and widespread effects of brine pollution are rarely recognized by most individuals, including those in state legislatures who formulate and pass into law the regulatory procedures. Unfortunately ground-water resources may be seriously and perhaps irreparably contaminated long before landowners are even aware that a problem exists. The water-bearing strata contaminated by brines may remain unusable, depending on the degree of contamination and on hydrologic conditions, for years, decades, or even millennia. In Texas, Alabama, Ohio, and probably many other oil-producing states, especially before enactment of protective controls, utilization of salt-water disposal pits caused the ground water to become so severely contaminated locally that in many instances the chloride concentration in the ground water was greater than that of the brines discharged into the disposal pits.

Contaminated Areas

An oil discovery well completed in Morrow County, Ohio, in 1961, excited a flurry of wildcatting and speculation throughout several areas in the central part of the State (fig. 1). About one of each three tests drilled for the next four years was a producer. The wells, which tapped porous reservoirs in the Copper Ridge Dolomite of Cambrian age, generally ranged in depth from 3000 to 3500 feet. Produced with the crude oil were large volumes of brine. Chloride concentra-

Reprinted with permission from *Ohio Journal of Science*, Vol. 71, no. 5, 1971, pp. 257-269.
*The Ohio State University, Columbus, Ohio.

tions of some of the brines ranged from 35,000 to 150,000 mg/l (milligrams per liter).

Unregulated disposal of these brines grossly contaminated several local shallow ground-water reservoirs. Before the enactment of specific legislation prohibiting such practices, much of the brine was removed from the well site by contract hauling. Contract tank truckers spread the waste on roads or indiscriminately dumped it into ditches, streams, swamps, quarries, or onto open or wooded fields. At other sites the brine waste was put into "evaporation" pits. A third and by far the most acceptable method was by means of a deep disposal well that returned the waste to the producing zone.

The most common disposal technique was by means of "evaporation pits"—a bulldozed hole in the ground in the vicinity of the well or separator. The pits, ranging from about 20 to 35 feet square, are three to 10 feet deep. Liquid wastes

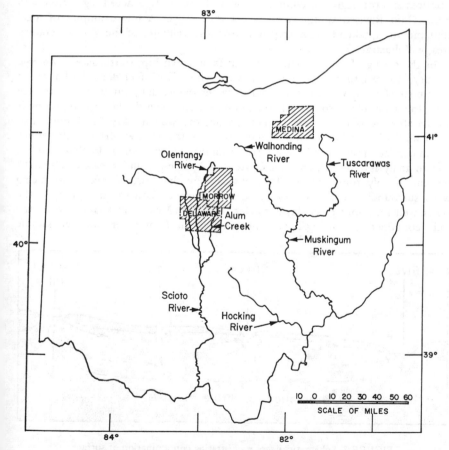

FIGURE 1. Location of brine contaminated areas on Ohio.

were, and in many places still are, dumped or pumped into the "evaporation" or, hydrologically speaking, "infiltration" pits. There the brine supposedly evaporates, but actually much infiltrates. These pits, of which perhaps hundreds still exist in Ohio, are now called "holding ponds"—but the change in nomenclature does not modify the way the procedure works nor the effects of this procedure on the ground water, that is, infiltration of brine from the pit into the ground. The pit-disposal technique has been one of the major causes of ground-water contamination in central Ohio. Moreover, abnormally high concentrations of chloride in many streams and rivers in the State, which are especially noticeable during periods of low flow, result from the natural discharge of highly contaminated ground water into the streams. A few "holding ponds" are conveniently drained by buried pipelines extending from the pond to a nearby stream. On the other hand, many of the ponds in areas of more permeable substrata are lined and permit only a minimal amount of leakage. Practices and the degree of compliance with the recently enacted regulations and rules (Ohio Revised Code) vary over a wide range. A few examples of local contamination, taken from a lengthy personal surveillance of the Morrow county area, will illustrate.

In the spring of 1967, a landowner in Peru Township, southeastern Morrow County, noticed that water from his well, which is 47 feet deep, had a strong salty taste (fig. 2). When it was used to water the garden, much of the produce died; the water began to corrode the plumbing, and even the family dog found it unpalatable. A water sample collected from the well in May 1967 contained 2800 mg/l of chloride, or more than 10 times the U.S. Public Health Service recommended limit of 250 mg/l for drinking water (1962). In September a second analysis showed a chloride content of 3300 mg/l, an increase of about 12 percent in only four months. The oil company that owned several producing wells surrounding the property apparently became alarmed, and pumped the water well for about three days "to clear it up." In May 1968 a sample from the well contained 5650 mg/l of chloride and in September another sample

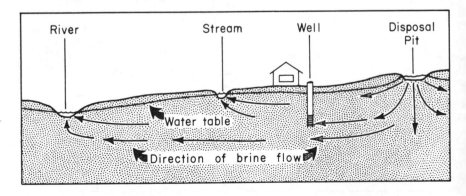

FIGURE 2. Schematic diagram illustrating contamination of surface water and ground water by pit disposal of oil-field brine.

contained 7600 mg/l. In November 1969 the chloride concentration was 7700 mg/l.

Several "evaporation" pits formerly existed in the vicinity of the contaminated well, one only about 600 feet away, in 1964 and part of 1965. Reportedly, these pits were all abandoned and filled by late 1965. Thus, it required one and a half to two years after the pits were abandoned for the brine to infiltrate through the ground and reach the well.

The well was not all that was affected by the brine from these pits. A small intermittent stream, a foot or two wide, which crosses the property, contained 40 to 61 mg/l of chloride in November 1969, and a major water course, nearly a mile away, contained more than 80 mg/l. In this part of Ohio, normal surface-water concentrations of chloride average less than 15 mg/l. These water courses were reflecting the chemical quality of the shallow underground water (fig. 2).

In early 1964, oil-drilling activities included an area in and around the village of Cardington in Morrow County. Several producing wells were drilled in the vicinity of the four shallow (30 feet) municipal water wells. Brines were discharged into pits or into an excavated ditch within 50 feet of these wells. Later in the year, village officials began to show some concern about the possibility of well-field contamination, and by January 1965 two of the village wells, which had formerly contained less than 10 mg/l of chloride, were contaminated with as much as 3750 mg/l. Eventually the entire well field had to be abandoned.

Elsewhere throughout at least 60 square miles in Morrow County, the effects of brine disposal began to appear, especially during 1964 and 1965. Several dozen areas, generally less than one square mile each, began to produce water containing, in some cases, several thousand milligrams per liter of chloride. Each area contained one or more producing oil wells and brine-disposal pits. Many domestic and stock wells had to be abandoned, and water was either hauled in or pumped from new and deeper wells at considerable cost. In many areas since 1967, however, the chloride concentration has decreased substantially; in other areas concentrations remain about the same or are higher.

Surface-water resources in Morrow County were also affected, generally to a lesser degree, due to three major causes: (1) dumping of brine directly into water courses, (2) intentional draining of evaporation pits into streams, and (3) natural discharge of polluted ground water into stream channels. In view of the stream pollution, the Ohio Department of Health began a surface-water surveillance program during the period April 1964—April 1966. Determination of chloride in samples collected weekly in Shaw and Whetstone Creeks showed chloride concentrations ranging from four to as much as 1350 mg/l. Concentrations were highest during August-December 1964, decreased until May 1965 (a very dry year), and then began to increase again. Even in mid-April 1966, when the surveillance program was terminated, concentrations at two of the three sites exceeded 110 mg/l, which is less than Public Helath Service limits but still more than seven times the normal background concentration. Surface-water samples taken at time of low flow in Whetstone, Alum, and Blacklick Creeks as late as

1967 still contained above-normal concentrations of chloride. Elsewhere in the basin, samples collected directly upstream from local ground-water contamination areas contained the least chloride; concentrations increased to a maximum in the contaminated areas and then slowly decreased downstream, due to dilution. Situations such as these existed as late as 1970. Unquestionably, however, the overall chloride content in both surface water and ground water in most of the areas contaminated during the early 1960's has decreased by several orders of magnitude.

In southeasterly parts of Ohio, such as the Hocking River basin, oil-field brines are heavily mineralized and may contain more than 320,000 mg/l of total dissolved solids. At places in this area, brines reportedly are stored in holding ponds to await some natural disaster. Commonly, heavy spring rains cause the pits to overflow, eroding the dirt retaining walls, which in turn allows the pit to empty on the ground or into a nearby water course. This effectively takes care of the waste through "an act of God." but it places a severe load on the streams' capacity to dilute contaminants. During midsummer 1962, the Ohio Water Pollution Control Board cited four Hocking County oil producers for brine contamination. In one case, an overflow stream from an "evaporation" pit killed all the vegetation for more than 350 feet from the pit. In another case, water samples collected from a stream above and below a pit contained 26,101 and 30,102 mg/l of chloride respectively, whereas, near another well, surface water samples contained 3,104 to 7,981 mg/l of chloride; a third pit polluted a stream with as much as 16,892 mg/l of chloride.

Delaware Area

The overall effects of surface- and ground-water contamination due to disposal of oil-field brines is readily evident from the preceding examples, but a more detailed examination of cause and effect indicates that the problem of contamination and natural cleaning is highly complex. A contaminated area in Delaware County, Ohio, has been described in considerable detail by Shaw (1966), Boster (1967), and Hulman (1969). Their work, techniques, and conclusions are summarized here in order to show not only the complexities of the problem, but also the considerable differences of opinions. Shaw, in his work, described the fluctuations in specific conductance, chloride concentration, and water level in a small area (about 20 acres) adjacent to the Olentangy River, near Delaware. Data were collected from 15 observation wells ranging in depth from five to 22 feet from July 1965 to December 1966. Boster had eight new observation wells constructed in the contaminated area, in which he measured the fluctuations in water level, chloride concentration, and specific conductivity, and attempted to determine the limits and degrees of contamination by means of electrical resistivity techniques. Hulman reinvestigated the Delaware area during the period January-October 1969 and found the chloride concentration greatly reduced, but still excessive at many times and in many wells.

The contaminated area lies northeast of Delaware on the flood plain of the Olentangy River. Here the floodplain, which is half a mile wide, is underlain by about 15 to 35 feet of alluvial and glacial gravels interbedded with sand, silt, and clay. The underlying bedrock, the Delaware Limestone of Devonian age, does not crop out in the area, but is exposed below the water surface in the present channel of the Olentangy River. However, shale has been reported by drillers; this shale, which is probably the Olentangy Shale of Devonian age, overlies the limestone and also crops out nearby. The land surface slopes gently toward the river, with an abrupt topographic break at the eastern margin of the floodplain. The southern part of the floodplain is slightly dissected by a small intermittent stream, Saunders Creek (fig. 3).

Precipitation in the Delaware area averages 38 inches per year. Flow in the Olentangy River is somewhat flashy, owing to controlled releases from nearby Delaware Reservoir and to peak periods of surface runoff. Much of the discharge of the river during later summer, fall, and winter is the result of effluent ground-water seepage.

During the period of major oil production in 1964–65, three oil wells were drilled in this area and four brine evaporation pits were created. Three of the pits were dug into the floodplain deposits and the fourth was excavated into the clayey till of the higher till plain to the east. The Skiles pit, which still exists, is adjacent to the Skiles no. 1 Hough oil well (fig. 3). About 300 feet to the south, across a section-line road and adjacent to the now-abandoned and plugged Slatzer No. 1 Ross well, were two of the pits, both of which are now filled (fig. 3). About 1100 feet to the east is the fourth pit, the Slatzer No. 2, which lies near the abandoned Slatzer No. 2 Ross oil well, and which is the only pit dug in till. This pit, like the Skiles pit, still exists. Shaw (1966, p. 71) reported that the Skiles pit was used during the period July 1, 1964, to July 30, 1965. The adjacent well produced about 126,000 barrels of brine during that interval and all of it was discharged into the pit. The brine averaged about 34,500 mg/l of chloride. The Saltzer No. 1 pit was used for at least 15 months and at least 110,000 barrels of brine containing 35,000 mg/l of chloride were pumped to it. Shaw calculated that the total amount of brine discharged in this area was equivalent to 1313 tons of chloride, or nearly 65 tons per acre throughout this salt-contaminated area.

Twenty-three shallow observation wells were drilled by Shaw and Boster in the area, with well screens at different altitudes. Unfortunately, geologic logs of the test holes are not available. Some of the screens were gravel-packed and some were not.

The lack both of geologic logs of the test holes and of information on well completion presents certain difficulties when attempting to interpret water-level, chloride, and conductivity data. The slow migration of a concentrated mass of brine down the water-table gradient is probably much more complex than is described here, and involves some interchange of water between the bedrock and the unconsolidated deposits. However, the hydrologic properties of the floodplain deposits, where the observation wells are located, are sufficiently distinct from those of the underlying bedrock and of the adjacent till to be treated here as a separate hydrologic unit.

The configuration of the water table as it existed in 1965 (Shaw, 1966), 1966

(Boster, 1967), or 1969 (Hulman, 1969), provides much interesting information. The maps prepared by these three investigators are similar; a modified version of one of Boster's (1967) maps is shown as Figure 3. The small southward-trending intermittent stream acts as a ground-water drain, as do Saunders Creek and the

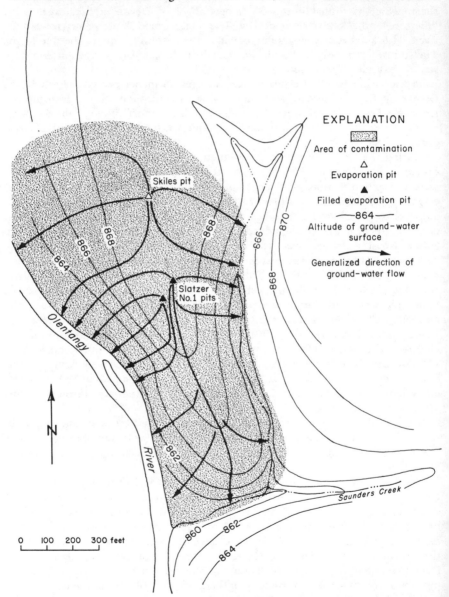

FIGURE 3. Water-level surface in the Delaware contaminated area in January 1966.

Olentangy River also. Ground-water flowlines, originating at the three evaporation pits on the floodplain, indicate that most of the contaminated water was discharged directly down the relatively steep ground-water gradient into the Olentangy River at the west boundary of the area, while a smaller, although still considerable, amount was discharged into the intermittent stream channels forming the eastern and southern limits. The discharge of briny ground water into these surface courses was expected because of the closeness (about 300 feet) of the pits to the streams. Evidence of such discharge is still visible, as for example, the abundance of dead trees and other vegetation between the pits and the stream channels. In addition, even during 1969 and 1970, surface-water measurements showed an increase in stream discharge and chloride concentration in the Olentangy River near the contaminated area.

There has been considerable decrease in chloride concentration in the ground water since July 1965, when the pits were abandoned, as the contaminated ground water continued to move to zones of discharge, showing that the aquifer is slowly purging itself of the contaminants. During 1965 and 1966, chloride concentrations in many wells near the disposal pits, although variable, were in excess of 15,000 mg/l. In 1969, chloride concentrations in these wells, with one exception, were less than 2,000 mg/l and generally were less than 1,000 mg/l.

Shaw (1966) believed that the steep ground-water gradient, coupled with the highly permeable nature of the floodplain deposits, would permit rapid ground-water exchange throughout the area. His estimate of the average velocity of ground-water flow here was about 1.5 feet per day. A study of the deposits, however, reveals that they are considerably less permeable than Shaw assumed. Although highly permeable gravel occurs near Saunders Creek, the deposits in many parts of the area include several feet of silt and clay, together with fine- to medium-grained sand. In addition, the deposits are highly lenticular along this reach of the Olentangy River. The second method used by Shaw to determine the ground-water velocity was based on the average chloride content, the area between adjacent isochlors (a contour line connecting points of equal concentration), and an estimate of aquifer porosity and saturated thickness. On the basis of these estimates and of other data, he calculated that the 20-acre area would be free of contamination within 16 months, or sometime between August and October, 1967—a prediction that is now known to be considerably in error. Probably one of Shaw's most significant errors was the assumption that the freshwater front would move uniformly through the sediments.

Boster (1967) pointed out that, on the basis of monthly measurements, the isochlors showed no appreciable change in configuration, but only in intensity. He recognized a gradual decline in chloride content throughout the area during his period of study, but believed that the decline was not linear with time. In fact, using some of Shaw's data, Boster calculated a chloride-free date of October, 1969. However, he also recognized the highly significant phenomenon of ion exchange, whereby sodium chloride is retained in soil owing to attachment of cations and anions to clay and silt. His final conclusion was that the floodplain deposits could return to their natural state by December, 1972.

Hulman (1969) began his study of the Delaware contaminated area during

January, 1969. Whereas Shaw and Boster were more concerned with time of dissipation of the contaminant, Hulman was more interested in the reasons for the continuous fluctuations in chloride values throughout the area. He reported that changes in chloride could be correlated with precipitation, that is, infiltration following periods of rainfall resulted in increased dissolved mineral concentrations in the ground water. He believed that salt crystals existed in the zone of aeration, having been formed during the infiltration of highly con-centrated brines in 1964 and 1965, which went into solution when water from infiltrated rainfall was available. He predicted that the area would remain contaminated until all of these salt crystals had been removed from the soil, and that complete removal might require decades.

A study of chloride fluctuations in specific wells compared with isochlor and precipitation data indicate the validity of Hulman's hypothesis. The positions of the 1,000—mg/l isochlor as it existed during selected months in 1965 and 1966 (traced directly from Shaw's maps) are shown in Figures 4, 5, and 6. It is interesting to note that the area enclosed by the 1,000—mg/l isochlor changes monthly and that the changes, even those from one month to the next, do not

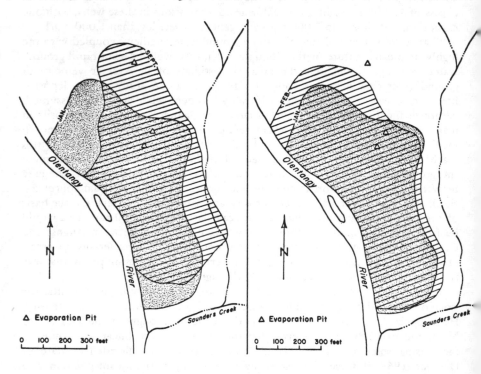

FIGURE 4. Areal extent of the contaminated area at Delaware enclosed by the 1000-mg/1 isochlor during September 1965 and January 1966.

FIGURE 5. Areal extent of the contaminated area at Delaware enclosed by the 1000-mg/1 isochlor during January and February 1966.

necessarily enclose smaller and smaller areas. This suggests that a linear relationship between chloride decrease and time does not exist. For example, the September 1965 1,000—mg/l isochlor encloses a larger area than does the January 1966 isochlor (fig. 4), indicating a natural flushing effect on the reservoir during the four-month period, but when the position of the January 1966 1,000—mg/l isochlor is compared with that of February 1966 (fig. 5), as well as with those of March and April (fig. 6), it is apparent that the contaminated area has increased in size. Significantly, both February and April were wetter than the previous months.

An examination of three similar maps presenting data for three specific months during 1969 shows similar phenomena (figs. 7, 8, and 9). The larger area enclosed by the March 500—mg/l isochlor was probably the result of 0.15 inches of rain that fell the day before the water samples were collected (fig. 7). A similar explanation (based on an antecedent rainfall) may account for the larger area enclosed by the September 500—mg/l isochlor (figs. 8 and 9). Monthly decreases in the area enclosed by the 500—mg/l isochlor are also readily explainable as resulting from less rainfall, as well as from natural cleansing of the

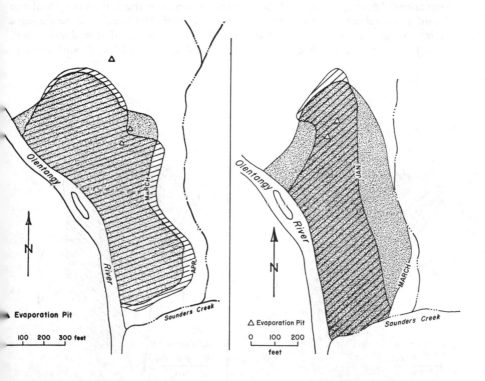

FIGURE 6. Areal extent of the contaminated area at Delaware enclosed by the 1000-mg/1 isochlor during March and April 1966.

FIGURE 7. Areal extent of the contaminated area at Delaware enclosed by the 500-mg/1 isochlor during January and March 1969.

floodplain deposits. It can be seen, then, that in this area, both the size of the contaminated area and the chloride concentrations are functions of precipitation.

Of all of the observation wells, a set of three, located about midway between the Skiles and Slatzer No. 1 pits, yielded the most useful data. None of these three wells, which are about two feet apart, were constructed alike (fig. 10). Well D—17s is shallow; the base of the gravel-packed screen is nine feet below the land surface. Observation well D—16s has a gravel-packed screen that bottoms 23 feet below the land surface, and the annular space above the gravel-packed screen was filled with clay to prevent vertical leakage along the pipe. Well D—3 is 23 feet deep, and the entire annular space from the screen to within two feet of the surface is gravel-packed. Well D—3 admits water from the entire saturated thickness: well D—17s admits water only from the upper part of the saturated zone; and well D—16s admits water only from the bottom part of the floodplain deposits. Graphs showing the variations through time of the chloride concentrations in these three wells are enlightening (fig. 11). In January, 1969, well D—16s, the deepest well, contained only 157 mg/l; well D—l7s, which is shallow, contained 547 mg/l; and well D—3, which is open to the full thickness of the aquifer, contained 286 mg/l. By the following April 1969, which came during a rainy period, a complete inversion of values had occurred—the deepest well (D—16s) contained 1,495 mg/l, the shallow well (D—17s) 837 mg/l, and the fully

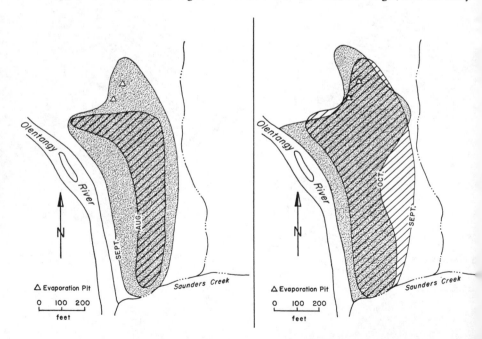

FIGURE 8. Areal extent of the contaminated area at Delaware enclosed by the 500-mg/1 isochlor during August and September 1969.

FIGURE 9. Areal extent of the contaminated area at Delaware enclosed by the 500-mg/1 isochlor during September and October 1969.

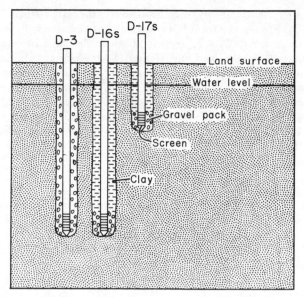

FIGURE 10. Completion details of observation wells D-3, D-16s, and D-17s.

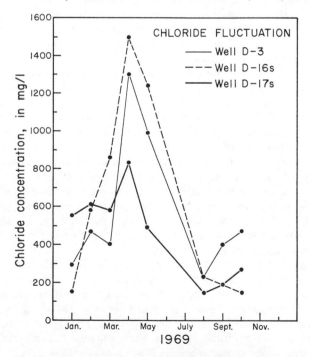

FIGURE 11. Fluctuation of chloride content in wells D-3,
D-16s, and D-17s during January-October 1969.

penetrating well (D—3), 1,305 mg/l. Evidently the salt, which is apparently chemically attached, in the form of crystals, to silt and clay held within pores in the upper part of the floodplain deposits, both in the zone of aeration and the zone of saturation, is dissolved and washed downward into the aquifer during periods of infiltration. If so, it is not possible to predict a time when the area will be completely clear of contamination, because it is entirely dependent upon the movement of soil moisture, and the rate, intensity, and duration of precipitation. Soil moisture, of course, too, depends not only upon the vagaries of nature, but also on the type of crops that might be planted in the contaminated area.

Other Chloride-Contamination Areas

Other areas in Ohio contain surface or ground water having high chloride contents. In some cases it is caused by man-made pollution, but in others the salt water occurs naturally. A few cases will be briefly described.

During the late 1940's, a local contamination problem occurred in north-eastern Ohio, in Medina County, due to the use of water flooding in the secondary recovery of petroleum, according to Stanley Norris, of the U.S. Geological Survey (written communication, 1970). Several hundred injection wells were drilled in an area of about 15 square miles. Water was pumped into the Mississippian Berea Sandstone at pressures as high as 400 psi to flush out oil from depths of about 400 feet. This resulted initially in contamination of a few private water supplies by saline fluids, forced into the shallower strata through abandoned and unplugged wells.

Kuhlman (1968) investigated the effects of street salting during winter on the quality of water in the Olentangy River in the vicinity of Columbus. He found a considerable chloride increase in the river immediately downstream from each storm-drain outfall. The combined high flows of both the river and the storm drains, however, served to rapidly dilute the salt-rich meltwater so that in no case did the chloride concentration exceed 100 mg/l.

Industrial wastes have become important sources of salt-water contamination in some parts of the State, particularly in the Tuscarawas and Muskingum River basins. These basins are potentially one of the most productive ground-water regions in the entire Ohio River basin. The greatest potential for ground-water development lies along the courses of the Walhonding, Tuscarawas, and Muskingum Rivers. Individual wells in gravel deposits along the floodplains may yield as much as 2,000 gallons per minute.

An example of this contamination occurs below Barberton, where large concentrations of chloride wastes dumped into the Tuscarawas River from a chemical complex present very serious water-quality problems. Because sand and gravel deposits along the valleys of the Tuscarawas and Muskingum Rivers form the most important aquifers in the drainage area, and heavy pumping induces recharge of river water, these aquifers will probably become more seriously contaminated, as development increases.

During low flow of the Tuscarawas River in September 1966, the content of dissolved solids (consisting largely of chloride) was 5,990 mg/l at Massilon, 2,800 mg/l below the Bolivar and Dover Reservoirs, and it was 2,360 mg/l even at Newcomerstown. The effect of the contamination persists at low flow in the Muskingum River, despite the dilution effects of the Walhonding River. Below Coshocton the content of dissolved solids in the Muskingum in 1966 was 1,500 mg/l, and even at McConnelsville dissolved solids were excessive at 1,090 mg/l.

Because of poor industrial waste-disposal practices, wells in southeastern Barberton have had to be abandoned because of contamination by salt water containing as much as 20,000 mg/l of chloride (Norris, 1955). Farther downstream, at Coshocton, the city well-field has had to be abandoned due to chloride contamination. At Zanesville, municipal wells in the sand and gravel of the Muskingum River floodplain produced water containing as much as 222 mg/l of chloride as early as 1950. The city water supply at Marietta, which is pumped from wells in the Muskingum floodplain, also has shown a marked increase in chloride.

Conclusions

Local contamination by the disposal of oil-field brines can result in serious economic hardships where domestic, municipal, or industrial wells are affected. In many areas streams may also be affected by the discharge of contaminated ground water. In those areas where precipitation is high and the strata are relatively permeable and near a drain, whether it be a well, spring, or stream, the contaminated water may be rapidly discharged, as is evidently true at Delaware. On the other hand, where localized deposits of sand or gravel comprise one or several aquifers within an area of otherwise low permeability, and under low hydraulic gradients, the natural cleansing of the aquifer may require decades, hundreds, or even thousands of years to complete. Considerations such as these are of profound significance in legal actions and will have important ramifications in the awarding of damages in ground-water-contamination suits. On the other hand, high concentrations of chloride in surface- or ground-water supplies in some cases may not be produced by contamination by oil-field brines, but may reflect natural conditions or disposal of other industrial wastes.

Acknowledgments

Appreciation for aid given during preparation of this manuscript is given to S.E. Norris, U.S. Geological Survey, and to R.L. Bates and Colin Bull, Department of Geology, The Ohio State University, all of whom read the report and offered several suggestions. Many data were obtained from reports of student research carried out in the Department of Geology, in particular reports by Jimmy Shaw, Ronald Boster, and Mark Kuhlman. Of special significance were the chemical analyses of water samples completed by Bruce Hulman, a graduate student at The Ohio State University.

References

1. Boster, R. S. 1967. A study of ground-water contamination due to oil-fields brines in Morrow and Delaware Counties, Ohio, with emphasis on detection utilizing electrical resistivity techniques. Unpub. M.Sc. thesis, The Ohio State Univ., 191 p.

2. Hulman, B. D. 1969. A ground-water contamination study in the Delaware area, Ohio. Unpub. senior thesis, The Ohio State Univ., 34 p.

3. Kuhlman, Mark. 1968. A study of chloride contamination in the Olentangy River in the vicinity of Columbus, Ohio. Unpub. senior thesis, The Ohio State Univ. 26 p.

4. Norris, S. E. 1955. The encroachment of salt water into fresh. p. 628. In Water, the year-book of agriculture 1955. U.S. Department of Agriculture.

5. Shaw, J. E. 1966. An investigation of ground-water contamination by oil-field brine disposal in Morrow and Delaware Counties, Ohio. Unpub. M.Sc. thesis, The Ohio State Univ., 127 p.

6. U.S. Department of Health, Education, and Welfare, Public Health Service. 1962. Drinking water standards. Public Health Service Pub. 956, 61 p.

Problems Arising from Ground Water Contamination by Sewage Lagoons at Tieton, Washington

R. H. Bogan*

This paper describes the unwitting contamination of several private water supplies near Tieton, Washington, a small farming community approximately 15 miles northwest of Yakima. The situation began with the conception, design, and construction of the town's sewage treatment facilities, and subsequently has led to a court of law and an acrimonious debate between local citizens and the community. The dispute involves alleged contamination of nearby ground waters by Tieton as a result of its sewage disposal operation.

The writer first became acquainted with the problem after it had become a matter of litigation. Factual data and information are limited. The early history of the situation, particularly that relating to design considerations and engineering study of the area, is vague and uncertain. Even under the most favorable circumstances it is often difficult to define accurately the character and extent of ground water contamination. Consequently, much of the discussion that follows is based on interpretation of local conditions in light of well-established general principles.

Sewage Disposal Operation

Tieton disposes of its sewage by means of a nonoverflow lagoon located approximately 0.6 miles south of town. The lagoon was first placed in operation in April 1957. It was designed to dispose of the domestic, commercial, and industrial wastes of the community through the combined effects of evaporation and percolation. Prior to 1957 the community disposed of its wastes by means of septic tanks. Thus domestic wastes have been admitted to local ground waters over a diffuse area for several years, but at a location 3,000 to 4,000 feet northwest of the lagoon. The lagoon actually serves, in effect, to concentrate the community's wastes at a new location.

Although the lagoon was intended to function as a nonoverflow operation, the remaining design criteria employed are for the most part unknown. Neither

Reprinted from *Public Health Service Technical Report* W61-5, 1961, pp. 83-87.
*University of Washington, Seattle, Washington.

the owner nor the engineer has described any material basis for design other than the assumption that the maximum percolation rate would be 0.25 inch per day (1). Similarly, it is not known what allowances, if any, were made for future service area expansion. Initially 124 households and 17 public, commercial, and industrial establishments were connected to the system. The present service area population is approximately 650. A typical rural community of this size might reasonably be expected to produce an average sewage flow of 50,000 to 60,000 gallons per day.

The lagoon as constructed consists of two cells having areas of 1.96 and 2.86 acres. One cell was intended to operate at depths ranging from 4 to 8 feet; the other was to operate at depths from 3 to 5 feet. Net evaporation in this area of Washington equals approximately 3.5 feet per year; this is equivalent to an annual average evaporation loss of about 3,200 gallons per acre per day. Clearly, much of the original sewage volume, together with soluble and colloidal constituents, must exit by infiltration into the surrounding ground. The Tieton sewage disposal operation is in reality a spreading basin or seepage pit and not a lagoon or stabilization pond in the ordinarily accepted sense.

On May 20, 1958, the average daily sewage flow was reported to be 130,000 gallons of which 50,000 gallons was described as domestic sewage and the balance as infiltration and industrial waste waters (1). With evaporative losses considered, an infiltration rate of about 0.9 inch per day over the entire 4.82 acres would be required to accommodate a flow of 130,000 gallons per day. The infiltration rate required to handle only the sewage flow component would be approximately 0.27 inch per day; this is very near the maximum infiltration rate of 0.25 inch per day reportedly employed as a basis for design. Evidently, sewage flows were in excess of those anticipated, and instead of the lagoon contents accumulating to the point of overflow, just the opposite occurred!

Initial infiltration rates as high as 15 inches per day were observed (1)! It was impossible to maintain water in either cell. Part of the difficulty was attributed to a leak in the lower dike. Even after the dike was repaired, however, it still was not possible to maintain water in both cells. Subsequently, sewage was admitted only to the cell into which the interceptor first empties. On May 20, 1958, more than 1 year after operation began, the infiltration rate was found to be approximately 3 inches per day (1). It was hoped that sewage solids would gradually seal the lagoon bottom, thereby decreasing the infiltration to a rate approaching the originally anticipated 0.25 inch per day. Apparently, the infiltration rate has decreased, for the water level after nearly 4 years of use has risen to approximately 3 feet.

Litigation

Two law suits were brought against Tieton for damages claimed as a result of invasion of ground water supplies by sewage in the vicinity of the Tieton lagoon. Both cases were tried before juries in the Superior Court of Yakima County. The first case, Pugsley vs. Tieton, came to trial in February 1959 and was concerned with pollution of a well approximately 250 feet south of and on property immediately adjacent to the Tieton lagoon. The second case, Cunningham et al.

vs Tieton, was based on the joint claims of 6 additional owners of property located for the most part within 1,500 feet of the lagoon; it was tried during October 1960.

In both cases the juries returned verdicts in favor of the plaintiff. For all except two of the parties in the Cunningham action, the jury found that the Tieton lagoon had adversely affected the ground water supplies of these people, at least to the extent that there was an element of doubt as to the potability of these waters. It was concluded that in their present condition and without some treatment these ground waters were no longer safe for human consumption.

The amount of damages awarded appeared to be based largely upon the influence ground water impairment had on the fair market value of the properties involved. In Pugsley vs Tieton, the plaintiff held that if the lagoon, by reason of odors or underground invasion of surrounding properties, deteriorated the value of adjacent properties there would be a taking or damaging within the meaning of the State of Washington Constitution (2). Both cases have been appealed and are pending before the Supreme Court of the State of Washington.

A second aspect of damage in the Pugsley case is that of a continuing nuisance, causing personal annoyance and inconvenience. In this regard, the plaintiff holds that a person can, as an item of damages wholly unconnected with the matter of depreciation in real estate value, recover for personal annoyance and inconvenience caused by the continuing nuisance resulting from the operation. The statute of limitations for nuisance is 2 years; hence, the plaintiff must initiate claims for damages every 2 years. Thus, it appears that Pugsley will continue to sue Tieton for nuisance damages at 2-year intervals or until the nuisance is abated or the operation ceases.

Discussion

Biological Considerations

Failure on the part of sanitary engineers and public health officials to accept diluted treated sewage as a legitimate domestic water supply is merely a reflection of the doubts and uncertainties currently held regarding the potability of such waters. This position may seem a trifle naive in light of the situation now prevalent in many inland drainage basins where today's sewage literally serves as tomorrow's water supply downstream. It must be recognized, however that where such conditions exist, these waters are subject to extensive treatment before consumption.

Perhaps the most serious and widely recognized hazard associated with domestic sewage is the possible presence of a number of pathogenic microorganisms. Recent evidence regarding the occurrence and persistence of viruses in domestic sewage adds still another element of doubt or reservation regarding the biological acceptability of treated and reclaimed sewages for human consumption (3).

Available evidence confirms what is intuitively obvious, namely that bacteria are quickly and effectively removed from sewage in passage through soil (4, 5).

TABLE 1. Presumptive coliform data—Pugsley well.[a]

Date	Result	Date	Result
1/ 4/54	-	8/14/57	+
1/18/57	-	10/ 2/57	+
3/15/57	-	12/ 5/57	-
6/21/57[b]	+	2/17/58	+
7/ 2/57	+	4/ 1/58	-
		5/21/58	-

[a]Results of bacterial examination conducted by the Yakima County Health District.

[b]Lagoon placed in operation in April 1957.

Unfortunately, a commonly held opinion that microorganisms rarely if ever penetrate more than 100 feet through continuous underground formations has assumed almost sacrosanct proportions. Obviously, the nature of the underground media through which ground water and sewage are free to move will determine the extent of bacterial penetrations. Circumstantial evidence was obtained by the Yakima County Health District (see Table 1) that indicates that E. coli traveled approximately 250 feet from the Tieton lagoon during the initial months of operation and entered a 160-foot-deep well.

Topography, location, and subsequent detection of anionic surfactants further confirm the conclusion that seepage from the lagoon was entering the Pugsley well.

Enteric viruses by their very nature should be free to travel considerable distances underground. It appears that many species of virus can remain viable outside the human body for several days and in some cases for months (3). At the present time, there is absolutely no evidence that the coliform bacteriological test can be relied upon to describe the persistence of enteric viruses in ground water. It is simply unrealistic to employ the coliform test as the sole criterion for judging the biological quality of ground waters known to contain sewage. Other things being equal, travel time and dilution appear to be the principal factors affecting virus penetration; however, in the absence of specific information, it is not possible to interpret the quantitative significance of either of these factors. Obviously, the greater the travel time and the greater the dilution, the less the chance of virus contamination. Other factors, such as adsorption and deactivation by constituents within the aquifer, may serve to remove viruses from sewage-contaminated ground waters; unfortunately, data are not available at this time that permit evaluation of such phenomena.

Geological Characteristics

Tieton is located in a long narrow valley that terminates in a mountainous area some 25 miles northwest of Yakima. The valley is formed by a series of nearly parallel basalt anticlines. A series of permeable sands and gravels of fluviatile and glaciofluviatile origin overlie much of the basalt bedrock. The exact thickness of the permeable surface formations is unknown, but, judged from wells in the area and from the general geological characteristics of the region, it ranges from 20 to 200 feet throughout the valley.

The valley floor is broken occasionally by basalt outcroppings. Thin clay lenses have been found throughout the surface formation. Prior to construction, three 5-foot-deep test holes were dug at the lagoon site; the top 3 feet were described as sandy loam and the next 2 feet as sandy loam and gravel (1).

Ground Water Movement and Zone of Influence

In the vicinity of Tieton, the valley is drained by the North Fork of Cowiche Creek. The Cowiche Creek drains into the Naches River about 2 miles west of its confluence with the Yakima River. Ground water movement in the valley tends to be in the same general direction as surface drainage, more or less straight down the valley. The general direction of flow may be altered in some areas by waters entering from neighboring hillsides and by discontinuities such as basalt outcroppings, etc.

Two attempts were made to determine the rate and direction of ground water movement below the Tieton lagoon. During May 1958, state and local health officials, together with a commercial laboratory, investigated ground water movement by means of chloride (Cl^-) measurements (6). In January 1959, the writer analysed samples of well water collected throughout the area for anionic synthetic detergent content. Even though the data obtained during these field studies cannot be viewed as incontrovertible evidence of sewage contamination, they are nonetheless indicative of general ground water flow patterns in the vicinity of the Tieton lagoon. Results of the detergent survey are shown in Figure 1. Travel times calculated from the Cl^- tracer study and the approximate zone of influence indicated by the detergent survey are shown in Figure 2.

Sewage leaving the lagoon apparently tends to form a shallow elongated mound of water resting on top of the normal ground water table. Results of the Cl^- tracer study indicate that velocities immediately below the lagoon are in excess of 300 feet per day! About 1,000 feet farther down the valley the velocity decreases to approximately 200 feet per day.

Infiltration rates encountered during the first year's operation, 3 to 15 inches per day, indicate that the ground in this area is exceedingly porous. If preconstruction soil test data are typical for the general area, then it seems reasonable to expect relatively rapid ground water movement, near that indicated by the Cl^- tracer study.

The zone of influence shown in Figure 2 is essentially a tentative evaluation of the extent and shape of the sewage mound in the vicinity of the lagoon. Some mixing and dilution of the sewage with ground water doubtlessly occurs down valley. Ultimately, a point will be reached where the residual effects or influence of the sewage are no longer significant. Just where the lower limits of ground water contamination lie in this case is unknown. Available information indicates that water originating in the Tieton sewage lagoon has reached rural ground water supplies as far as 1,500 to 2,000 feet down the valley in approximately 6 days.

In retrospect, it appears that little if any consideration was given to the

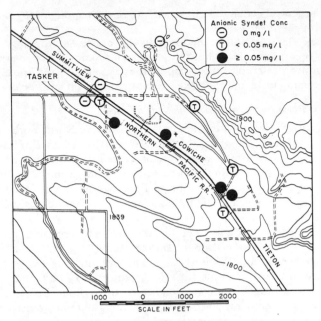

FIGURE 1. Anionic synthetic detergents in down-valley well waters define ground water movement and probable extent of sewage-induced ground water mound.

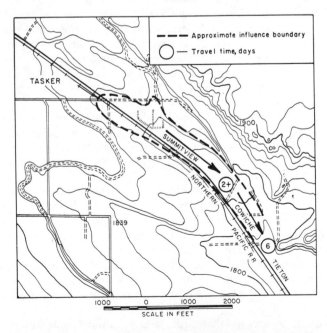

FIGURE 2. Tieton ground water zone of influence and travel times.

influence of sewage infiltration on ground water quality. Indeed, it seems that even the infiltration characteristics of porous surface formations at the lagoon site were poorly understood. In principle, there is little difference between the conventional or more common practice of discharging wastes to surface waters and that of ground disposal. In either case, the waste disposal operation must be based on a thorough understanding of the self-purification characteristics of the receiving water mass and the effect of such action on subsequent water users.

References

1. Answers to interrogatories to the defendant, *Pugsley v. Tieton,* No. 42005, Superior Court of the State of Washington in and for Yakima County.
2. Private communication from Blaine Hopp Jr., Tonkoff, Holst & Hopp, Attorneys at Law.
3. Clarke, N. A. and S. L. Chang. 1959. Enteric viruses in sewage. Amer. Water Works Assoc. 51, 1299.
4. Butler, R. G., G. T. Orlob and P. H. McGauhey. 1954. Underground movement of bacterial and chemical pollutants, J. Amer. Water Works Assoc. 46; 97.
5. Krone, R. B., P. H. McGauhey, and H. B. Gotaas. 1957. Direct recharge of ground water with sewage effluents. Proceedings ASCE, Paper 1335, p. 83.
6. Unpublished report by G. H. Hanson, District Engineer, Washington State Pollution Control Commission.

Infectious Hepatitis Outbreak in Posen, Michigan

J. E. Vogt*

During the spring and early summer of 1959, an outbreak of infectious hepatitis occurred in the village of Posen, Michigan. Posen is a Polish Catholic community with an estimated population of 400. It is located in Presque Isle County in the northeast part of Michigan's lower peninsula.

The people in Posen as a whole form a very close community. An occasional trip to Rogers City or Alpena, the nearest communities of any size, is the extent of their normal travel. Community life centers around the church, the parochial and public schools, the Chamber of Commerce building, where wedding receptions, showers, and other social activities are held almost every weekend, and the local theater.

Geology and Hydrology

Geologically the area is very interesting. In the immediate area around Posen there is a thin veneer of glacial till, with a maximum thickness of 3 feet, overlying the bedrock. Outcroppings of rock are numerous throughout the area. The bedrock formations belong to the Traverse Group of the Devonian Age and consist mainly of limestones with some shale beds. These formations dip toward the center of the state at about 40 feet per mile.

No topographic maps of the Posen area are available, and without them, interpretation of the hydrology of the area is difficult. By use of the information available, however, the following general conclusions can be drawn regarding the hydrology.

A topographic high apparently extends through the Posen area in a north-west-southeast direction. This could be caused by some resistant limestone beds, within the Traverse Group, forming a low ridge paralleling the strike of the rock formations. The relief is not very great, but it is enough to be a controlling factor in the surface drainage. South of a northwest-southeast line through Posen, the streams drain southward, and north of this line they drain northward.

Water level measurements in wells indicate that Posen is situated on a ground

Reprinted from *Public Health Service Technical Report W61-5, 1961, pp. 87-91.*

*When this report was written, J. E. Vogt was employed by the Michigan Department of Health, Lansing, Michigan.

water divide that trends in a northwest-southeast direction. This ground water divide conforms to the surface drainage pattern, and parallels the strike of the rock formations in the area. South of Posen, the ground water gradient is toward the southwest, and north of Posen, it is toward the north and northeast.

Water Supply

Posen has no public water supply and all the individual wells are drilled into the bedrock, where water is obtained from joints, fractures, and partings along the bedding planes of the limestone formations. The wells range in depth from 11 to 180 feet. The most usual depth ranges apparently are 30 to 40 feet, 60 to 70 feet, and 90 to 100 feet. Part of this wide range in the depth of wells is due to differences in elevation of the ground surface, and some of the deeper wells were probably attempts to obtain a "safe" water supply.

No relationship was apparent between the depth of wells and the quality of water. The observation, at one point in the investigations, was advanced that the majority of the contaminated wells were from 50 to 65 feet deep. When the data were compiled later, it was found that there was no correlation between depth of well and water safety.

Well Construction

A rather unique method of well construction is used in the area. All wells are 6 inches in diameter with casings that terminate at the ground surface. Most of the wells are cased to only a shallow depth, generally just through the glacial till and into the rock a short distance. The lengths of the casings range from 10 to 30 feet; and one well has no casing at all. Water was entering one well, for which the casing terminated at a shallow depth, from a shallow formation above the static water level and below the casing.

In many installations the casings were cut off at the ground surface. Hand pumps were then installed on top of the casings, to be used during power outages. In practically all wells, however, no seal was installed between the hand pump drop pipe and the well casing, leaving the casing open for surface drainage. Where there was no hand pump, the well might be covered with some loose boards. Furthermore, no attempt was made to seal the bottom of the casing in the rock.

The power pumps were generally located in the basements of the homes with a suction line running out to the well casing. The portions of the construction features that were visible and conversations with the residents left little doubt that often the connection between the suction line and the well casing was not watertight. Accordingly, numerous opportunities for contamination resulted from the poor construction features—open top casings, unprotected suction lines, unsealed connections between suction lines and casings, and no seal between the bottom of the casings and the rock.

Location

As for location, most of the wells were too near septic tanks, tile fields, or seepage pits. This is particularly true when the thin veneer of soil on top of the limestone is considered. The excavation for the septic tank sometimes had been blasted out of the rock and the tile field laid in filled ground above the rock. Little doubt existed that septic tank effluent percolated vertically through the thin mantle of drift into the rock and then moved laterally for great distances, with little change in the characteristics of the effluent.

Water Quality

Laboratory tests of water samples from the wells substantiated the field observations. Thirty percent of the samples collected over the previous 3 or 4 years by the local health department had shown evidence of contamination. When the hepatitis cases began to appear, the well waters in the community were suspicioned as agents in the spread of the disease. During the investigation of the village wells a rather complete survey was made of all wells north of the corners. The survey was concentrated on the north side of the village, since practically no cases had appeared south of the main square of the village. As much information as possible was obtained about the construction of the wells and their location with respect to sources of contamination. Bacteriological analyses were made on all samples, and nitrates, nitrites, and detergents were checked on many. Within the village 47 percent of the samples showed the presence of coliform organisms. Forty-four percent of the wells were inadequately located with respect to sources of sewage pollution, and 70 percent had buried unprotected suction lines. Of the 10 wells checked for detergents, four were positive. One of these wells served the Chamber of Commerce building, which was frequently used for community affairs such as wedding receptions. A second well served the residence of the local druggist, and all the members of that family had suffered severe cases of hepatitis. The other two wells will be discussed in more detail later.

Weather Effects

The area had experienced subfreezing weather since Thanksgiving. For 3 consecutive days near the middle of March, however, temperatures rose to well above freezing and substantial melting of snow resulted. April 3 the village was blanketed by a 10-inch snowfall, followed by light rain. By April 15, all the snow was gone and runoff was complete. Wells in the village were producing a highly turbid water, which might be expected in view of the geology, the construction of the wells, and the heavy runoff. The local health department advised all residents to boil their water. Conditions were aggravated further during the first week in May by heavy rains; practically every basement in the community was flooded.

Spread of the Disease

Infectious hepatitis was reported first in the nearby community of Hillman in January 1959. The first cases appeared in Posen in the middle of April, apparently unrelated to those in Hillman. The "explosion" occurred in three families living in "Upper Posen," the north side of the community. About the middle of March one of the families was visited by a relative who, during his stay, became ill. The illness was diagnosed as infectious hepatitis. This visit coincided with the above-freezing temperatures that produced considerable runoff, later compounded by heavy snow and light rain.

The septic tank serving the home with the first hepatitis case was only 6 feet from the well. This well, like most of those in the community, had a 6-inch casing, driven to the shallow rock formation and cut off flush at the ground surface. A power pump was located in the basement and was connected to the well by a buried suction line. A hand pump was set over the casing, but was not sealed to provide a watertight joint. It was easy to understand how this well, so near the septic tank and so poorly constructed, had become contaminated.

The two houses immediately south of the one in which the first hepatitis case developed were served by one well, which was located only 10 feet from the septic tank on the first property. This well also was pumped through a suction line from the basement of the house. The top of the casing was "protected" by a couple of loose-fitting boards.

Within a 3-day period about 4 weeks after discovery of the first case of infectious hepatitis, 16 cases appeared in the three families. From this nucleus hepatitis spread rapidly through the community at epidemic proportions. Eighty-nine cases were reported; an epidemiological study showed later, however, that many cases were not reported. When one person in a household was diagnosed as having hepatitis, others in the household with similar symptoms were merely put to bed and no report made. Furthermore, it is very likely that many subclinical cases were not reported.

Hydrology a Factor

Superimposed upon the ground water divide that runs through Posen is what appears to be a drawdown cone that reverses the natural flow of ground water. This results in ground water flowing into Posen from all directions, toward the low point of the cone of depression near the north central part of town. The existence of this drawdown cone is supported by the fact that numerous wells around the perimeter of the town yielded bacteriologically safe samples, whereas the wells within the village yielded a large percentage of unsafe samples.

As the ground water moves down the drawdown cone toward the center of the village, the fringe area wells intercept uncontaminated water. As the ground water continues down the cone, contamination from septic tanks, privies, sink drains, etc. is added. Since the ground water moves through fractures, partings along the bedding planes, and other openings in the limestone, little or no straining or filtration takes place and wells within the village intercept contaminated water.

The drawdown cone near the center of the village could be caused by either of two things. If the permeability of the aquifer in the Posen area is low, the combined pumpage from the concentration of domestic wells in the town may have deprived water levels locally and created the drawdown cone. The low permeability of the aquifer has since been verified by an automatic recording device that has been recording ground water elevations in a well near the center of town. The recorder showed daily fluctuations of 1 to 1.5 feet and seasonal changes from 6.5 to 10 feet. This is significant, since only low-capacity domestic pumps are used in this area.

The second possible explanation for the drawdown cone is that in the town there is a deep uncased well that penetrates a water-bearing formation in the limestone that has an artesian pressure surface lower than the water level in the upper formations. Ground water, under these conditions, would migrate from the upper water-bearing formations down the well bore into the lower water-bearing formation. This would create a drawdown cone in the upper formation from which the wells in the town extract their supply.

A well more than 300 feet deep is reported to exist in the village; however, this well was never located. A 300-foot well in Posen would probably penetrate the Rogers City limestone. The 80-foot thickness of Bell shale overlying the Rogers City limestone would constitute an effective aquiclude, and the artesian pressure surface in the Rogers City limestone could be lower than the water level in the Traverse formations. An uncased well that passes through the Traverse limestone and the Bell shale and penetrates the Rogers City limestone might account for the drawdown cone that apparently exists in the water table of the shallower Traverse limestones.

Corrective Measures

This discussion certainly has indicated that Posen has a water supply problem. Correction of the problem on an individual basis would be nearly impossible. The small size of the lots and the innumerable sources of contamination make adequate isolation of wells virtually impossible. Also, isolation is impractical because of the lack of natural purification of water in the area as it travels through the limestone.

The best solution to the water problem in Posen probably is the construction of a municipal water supply. Great care would be needed in the selection of the proper site for any municipal wells. First consideration probably would be given to the area on the south side of the village. Posen officials have had an engineering study made and a report prepared on the construction of a municipal water system; however, little further progress has been made, since any project would be costly because of the rock excavation necessary in the construction of a distribution system.

A bill introduced in the current session of the Michigan Legislature is designed to prevent situations like that in Posen. The bill provides for the licensing of well drillers by the State Health Commissioner and issuance of permits before wells are drilled.

Summary

An outbreak of infectious hepatitis occurred in the northern part of Posen, Michigan, in 1959. Epidemiological studies indicated rather clearly that the first

16 cases were water-borne. The outbreak progressed rapidly south of the initial "explosion" and continued toward the center of the village where the progress of the disease was sharply curtailed.

The pattern of the disease conformed to a developed theory of the hydrology of the area. The virus apparently was introduced into the ground water through septic tank effluent on the north side of the community. According to the theory, melting snow and spring rains then flushed the virus through fractures in the limestone to the water table where it spread laterally. The virus moved southward down an inverted cone to the north central part of the town, infecting wells along the way. After reaching the low point in the drawdown cone, the virus could travel no further, probably accounting in part for the rather sudden decrease in the number of cases south of the center of the village.

Acknowledgment

The author wishes to express his appreciation to Norman Papsdorf, Sanitary Engineer, Michigan Department of Health, for his field study and to L. David Johnson, Hydrogeologist, Geological Survey Division, Michigan Department of Conservation, for his assistance in evaluating the hydrology of the area.

Good Coffee Water Needs Body

Wayne A. Pettyjohn*

Crosby is a small village, formerly a railhead center, in the northwestern part of North Dakota and a few miles south of the International Boundary. As in many places in North Dakota, water is not in abundant supply. When Crosby was first established, surface water was unavailable, and dug wells were used to supply homesteaders and the community at large. Most of these hand-dug wells, commonly several feet in diameter, were relatively shallow. They were replaced in recent years with a municipal water system supplied by two drilled wells more than 150 feet deep.

In the central part of town, however, there remains an old large-diameter dug well, about 38 feet deep, covered by boards and concrete. Water was formerly withdrawn from the well by a handpump and used by a great number of people who did not like the taste of the water from the city's deep-well supply. Water from the dug well was used specifically in making coffee; it reportedly produced "the best coffee in the State."

During a ground-water investigation of the Crosby area by the U.S. Geological Survey in the early 1960s, a sample of water was collected from the dug well and analyzed (Armstrong, 1965, p. 111). The water contained the following constituents:

Sulfate	846 mg/1
Chloride	164 mg/1
Nitrate	150 mg/1
Dissolved solids	2176 mg/1
Hardness	1300 mg/1

It is evident that this water is highly mineralized. It contains more than four times the limit of 500 mg/l of dissolved solids recommended by the Public Health Service (1962). The local people referred to it as having "body."

Actually, the water had more body than these people realized. The high

*The Ohio State University, Columbus, Ohio.

concentration of nitrate, which far exceeds the Public Health Service limit of 45 mg/l, coupled with the higher-than-normal concentration of chloride, suggested that the well was contaminated by sewage wastes. The apparent contamination was brought to the attention of the local city health official, who declared the well unsafe and removed the pump handle. Immediately many people became angry because their supply of good coffee water had been terminated, apparently for no good reason. After all, they had been drinking the water for years, and no one had ever died from it, or even gotten sick—as far as they knew. Nevertheless, the pump handle was not reinstalled.

Some of the old timers reported that this ancient well had been dug near the site of a former livery stable, which had been built sometime in the late 1800s and had been operated until about 1930. Waste products from the horses evidently contaminated the ground-water supply. These wastes also provided the "body" that patrons of Crosby's dug well prized so highly.

E. C. Wood (1962) described an interesting case of ground-water contamination caused by the infiltration of waste materials from an ancient gasworks in Norwich, England. In the early 1950s, a 36-inch diameter well was drilled 150 feet into chalk deposits underlying the city. Although water from the well was of acceptable quality, both biologically and chemically, the yield was insufficient, and an adit was driven horizontally from the bottom of the well into the chalk in an attempt to provide more water.

After construction of the adit, the water was sampled and found to contain phenols in the amount of about 0.2 mg/l, causing the well to be abandoned. (The U.S. Public Health Service has recommended a phenol limit of 0.001 mg/l, since at higher concentrations this substance creates significant taste problems.) Mr. Wood, his curiosity aroused, entered the well by means of a bucket and upon examining the roof of the adit found several zones seeping black tarry material. Analyses of this material showed it to consist of tarry carbon with a trace of phenols and volatile matter.

Although the well site was several miles from an existing gasworks plant, the composition of the tarry materials suggested that they had originated from the distillation of coal or similar material. Further investigation brought to light the fact that the first gasworks plant in the city had been constructed around 1815 at the exact site of the contaminated well. In March 1830 the original plant was abandoned. Thus, Wood concluded, the tarry material that contaminated the well had been present in the chalk for at least 120 years.

Perhaps one of the most unique cases of ground-water pollution was provided by G. J. C. Nash (1962) during a discussion of Wood's report. Nash stated that officials of a gasworks plant on the south coast of England once claimed that the presence of hydrogen sulfide in the plant well was due to drainage from a Black Plague burial pit. Apparently, the well was bored through a seventeenth-century graveyard.

In 1925 Boy Scouts in the central Ohio city of Delaware conducted "an outhouse" survey (Bill Rice, oral communications, March 1972). The purpose of the investigation was to determine the distance between each homeowner's outhouse and his well, from which a rough estimation could be made of whether the water supply was biologically safe. The data were given to the local health agent, and principal investigator, Dr. B. F. Higley.

Dr. Higley was concerned about the relation of well water to public health in Delaware because the town was constructed on a nearly flat, till-covered area. Shallow sandy zones, a few inches to a few feet thick, served as a source of water to wells. Although the sandy zones were of limited areal extent, it was possible and in fact highly probable that some of these permeable layers were acting both as a source of drinking water and as a receptacle for human wastes, thus forming an early and rather primitive example of water recycling and reuse.

One homeowner who had built a privy only a few feet from his shallow well remained unconvinced by Dr. Higley's most eloquent arguments and would neither dig a new well nor move the outhouse. To prove that there might be a hydrologic connection between the privy and the well, Dr. Higley poured five to ten pounds of salt into the privy. Within two weeks the well water began to taste salty and the homeowner threatened to sue the good doctor for contaminating his well. Ultimately, a new well was drilled on the other side of the house.

In early 1961, members of the Division of Water, Ohio Department of Natural Resources, investigated an unusual, but by no means unique, water pollution problem in the northeast Ohio community of Bellevue (Ohio Division of Water, 1961). Contamination of a highly-permeable, limestone aquifer had resulted from the dumping of household, municipal, and industrial wastes into scores of sink holes and drilled wells. In many instances septic tanks were used, but overflow from the tanks was allowed to discharge into wells. The sewage effluent contaminated the ground water in an area 5 miles wide and more than 15 miles long as it slowly moved down the water-table gradient toward Lake Erie (fig. 1).

Bellevue's first municipal water supply was from a surface reservoir constructed in 1872. By 1919 the practice of disposing of sewage into disposal wells and sink holes was already well established and many wells had been contaminated. To augment the surface water supply, municipal wells were drilled in 1932 outside the known area of contamination. Following a major flood in June 1937, caused by ten inches of rain during a three-day period, many sink holes overflowed, pouring raw sewage on the ground throughout several square miles and contaminating many more wells and cisterns. Additional water-supply wells were drilled by the city in the early 1940s. The average yield per well was 500 gpm (gallons per minute) from depths ranging from 137 to 200 feet. By 1944 several of the municipal wells were contaminated and plans to build a soybean processing plant were abandoned because a well, 230 feet deep, drilled at the prospective plant site was contaminated. All industrial and municipal water wells were abandoned by 1946. On 4 July 1969, the city was again inundated by heavy runoff caused by ten inches of rain falling over a 16-hour period. The city's main street was covered by 16 feet of water. Infiltration filled the underground reservoir and the waste poured over the ground throughout the city and adjacent areas. About 1,100 basements were flooded and surface water supplies were contaminated. Plans to construct a sewage treatment plant were shelved year after year due to "excessive" cost. Only in 1971 was a sewage treatment plant finally completed with federal assistance.

In the decades during which waste materials were dumped into drilled wells in and around the city, wells occasionally would become plugged with sewage, necessitating redrilling. With the increased use of detergents, plugging ceased to be much of a problem. Division of Water investigators found more than 1,500 disposal wells within Bellevue's city limits—an area less than four square miles (fig. 1). Of 32 samples of ground water collected in the vicinity of Bellevue, 27 contained ammonia. Detergents (alkyl benzene sulfonate or ABS) were found in 22 samples, and all contained nitrate and phosphate. Examples of concentrations found in samples collected in February, March, and April 1961 are shown below and reported in mg/l:

Property owner:	Date of collection:	ABS	Ammonia as NH_4	NO_2	NO_3
Weiland	2-6-61	0.1	0.1	0.5	29
	4-18-61	0.1	0.2	0.05	52
Thomas	3-7-61	0.1	0.2	0.00	92
	4-18-61	0.3	0.1	.00	158
Neill	3-7-61	0.1	0.2	1.0	41
	4-18-61	0.1	0.1	.05	38
Andrews	2-6-61	0.1	0.2	1.5	26
	4-18-61	0.2	0.2	.00	36
Adams	3-5-61	0.2	0.1	0.05	77
	4-18-61	0.2	0.3	0.20	91

The ground-water resources in the Bellevue area, and in areas downgradient from the town, are obviously grossly contaminated and have been for more than a half century. The Division of Water report states: "Stories have been related to us during this investigation of wells which yielded easily recognizable raw sewage (including toilet tissue) while being drilled. Others have foamed because of high detergent content, and, still others, the contents of which are best left to the reader's imagination!"

The attitude of some of the local officials, however, was rather curious, at least during 1961. One official stated the cry of ground-water pollution has "been made every three or four years for the last 20 years, and has never been proven" (*Sandusky Register* 28 June 1961). A commissioner of the Bellevue Health Department said that city officials, since the original inception of the use of sink holes, had made studies of the area and could find no evidence of water contamination (*Fremont News*, 24 June 1961). Bellevue's city engineer reported that the city had wells, which it used for drinking water, and the water was not contaminated. However, an examination of water from two of these wells proved to be rather enlightening. The location of the wells and the poor condition of the casings had led the Ohio Department of Health to speculate

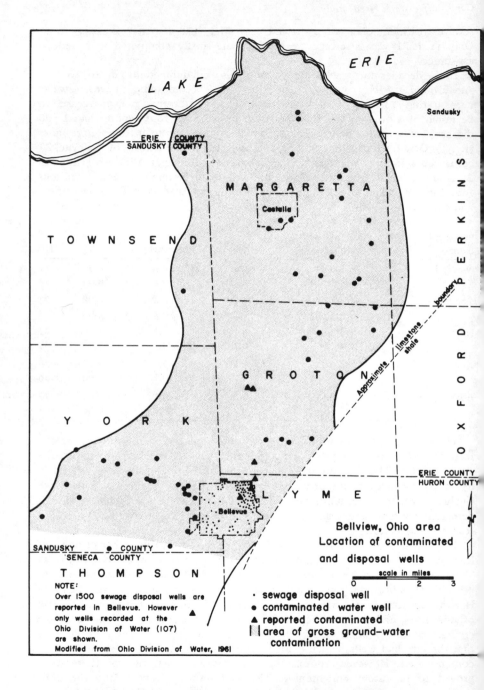

FIGURE 1. Location of contaminated and disposal wells in the Bellevue, Ohio, area.

that these wells probably could not be approved as a source of public water supply. Chemical analyses showed one of the wells (212 feet in depth) to be grossly contaminated; the chemist described the water as "foul." Fortunately, neither of these wells had been used since 1946.

Literally hundreds of ground-water contamination cases, such as those described herein, are to be found in the literature. All of them tend to point out the fact that inadequate disposal of waste materials can result in contamination, even though the contaminants have been abandoned and forgotten for decades.

Of all that was done in the past, you eat the fruit, either rotten or ripe.

—T. S. Eliot

References

1. Armstrong, C. A. 1965. Geology and ground-water resources of Divide County, North Dakota, part 2, Ground-water basic data. N. Dak. State Water Comm., County Ground Water Studies 6, 112 p.
2. Nash, G. J. C. 1962, Discussion of paper by E. C. Wood. *In* Proc. Soc. Water Treatment and Examination 11; 33.
3. Ohio Division of Water. 1961. Contamination of underground water in the Bellevue area. Ohio Dept. Nat. Resources mimeo. rept., 28 p.
4. Wood, E. C. 1962. Pollution of ground water by gasworks waste. *In* Proc. Soc. Water Teatment and Examination 11:32-33.
5. U.S. Department of Health, Education, and Welfare, Public Health Service. 1962. Drinking water standards. Public Health Service Pub. 956, 61 p.

The Movement of Bacteria and Viruses Through Porous Media

John C. Romero*

The Colorado Board of Examiners of Water Well Drilling and Pump Installation Contractors was established in early 1968. By October the Board adapted a set of rules and regulations which provide minimum standards for location, construction, modification or repair of pumping equipment.

The Board's current concern is with the sanitary quality of those waters which are to be used for domestic purposes; in particular . . . human consumption. Realizing that proper sanitary protection of domestic wells involves the prevention of the intake of contaminants, the Board of Examiners introduced a set of standards which provides for the location of wells and well casings with respect to sources of contamination. The standards were chosen arbitrarily and do not consider the differences between chemical and biological contaminants.

It is the purpose of this report to briefly point out some of the past and current developments regarding the nature of ground-water pollution and, in particular, data related to studies of the length of travel of bacteria and virus-laden water when injected on both soil surfaces and in water wells.

Two distinctly different aspects of pollution travel are considered. They are: (1) the movement of bacteria and viruses *downward* with percolating water and (2) the lateral movement of identical pollutants once they have reached the zone of saturation.

Infiltration of Contaminants by Surface Spreading

The spreading of sewage-plant effluents as a means of recharging the ground-water reservoir was first studied nearly 40 years ago. It was found that reclaimed water was satisfactory for irrigation and stock purposes. Since that time a large number of schools and private organizations have studied such a means of water reclamation. Resulting data indicates a tremendous potential for water

Reprinted from *Journal of Ground Water*, Vol. 8, no. 2, 1970, pp. 37-48, with permission of the author and the National Water Well Association.

*Division of Water Resources, Department of Natural Resources, Denver, Colorado.

reclamation because if the sewage effluent and surface material used as the spreading basin are controlled, most if not all of the water's remaining harmful pollutants can be eliminated.

It is now accepted that great numbers of bacteria are effectively removed by percolation through a few feet of fine sand. The removal process involves mechanical and biological straining (a result of soil clogging) and/or death induced by environmental changes (soil defense mechanisms).

Soil clogging, whether it be due to physical or biological processes, is an increase in physical resistance to flow resulting from changed friction or velocity coefficients or in reduction in size and volume of pore spaces. Generally soil clogging occurs within the first six inches of soil. "Soil defense" mechanisms involve the destruction of harmful bacteria by: (1) the latter's inability to adjust to abrupt temperature changes; (2) "oxygenation" and "nitrification"; and (3) destruction by pre-existing soil bacteria.

The following summaries were obtained from various articles dealing with surface spreading of pollutants.

I. University of California Sanitary Engineering Research Laboratory at Richmond, California

Studies of surface spreading at Whittier and Azusa, California (in SERL Report 67-11, pp. 70) led to the conclusion that water of a bacterial quality suitable for drinking purposes can be obtained by passage through a minimum of 3 to 7 feet of soil even though it might be quite coarse. At Whittier secondary sewage effluent containing a coliform density exceeding 110,000 per 100 ml produced a concentration of coliforms of 40,000 per 100 ml at a 3-foot depth in 12 days. NONE appeared at *greater* depths even after prolonged spreading. At Azusa similar effluent carrying 120,000 coliforms per 100 ml produced a density of 6,000 per 100 ml at depths of both 2.5 and 7.0 feet. When primary sewage was applied the 7-foot coliform count rose to 1,200,000 per 100 ml. Subsequent experiments on beach sand showed coliforms reduced from 70,000 to 700 after 12 feet of travel.

Studies with infiltration ponds at Lodi, California indicated that at some point between the 4 and 7-foot levels the coliform count declined to less than the USPHS minimum standard of 1 organism per 100 ml. An additional fact stemming from the Lodi experiments is that the number of coliforms penetrating one foot or more is essentially independent of the intensity of pollution of the waste water spread.

Subsequent lysimeter studies of several other soils in the Lodi area revealed a very pronounced removal of organisms in the top 0.5 to 1.0 cm of soil AND in a secondary "limiting" zone 50 to 60 cm (about 22 inches) below the surface. Soil conditions and coliform count at a depth of three feet are given in Table 1 and the change in coliform concentration with depth for Hanford sandy loam is illustrated in Figure 1. The initial coliform count was 1.9×10^8 coli per 100 ml.

One of the most recent and thorough studies of reclamation of water from sewage effluents is the well-known Santee Project near San Diego, California.

TABLE 1. Soil textures and coliform count at a depth
of three feet for five California soils.

Name	Effective Size	Sewage Infiltration Rate (ft per day)	Coliform mpn/ml
Hanford sandy loam	0.0056	0.3	45
Hesperia sandy loam	0.0020	0.2	45
Columbia sandy loam	0.0034	0.3	45
Yolo sandy loam	0.0155	0.3	24,000
Oakley sand	0.015	0.1	2,400

TABLE 2. Summary of bacteriological analyses
(Coliform, Fecal Coliform and Fecal Streptococci).
All results are MPN/100 ml.

Sampling Station	Oxidation Pond	200-ft Well	400-ft Well	Interceptor Trench
Coliform Bacteria				
No. of Samples	29	29	29	28
Maximum	11 Mᵃ	3,300	35,000	22,000
Minimum	330,000	<180	78	20
Median	1.1 M	780	2,800	560
Fecal Coliform				
No. of Samples	29	29	29	28
Maximum	1.3 M	200	3,300	110
Minimum	4,500	< 18	< 18	<18
Median	210,000	<180	<180	<18
Fecal Streptococci				
No. of Samples	29	29	30	29
Maximum	490,000	54,000ᵇ	9,200ᵇ	230
Minimum	450	2	2	< 1.8
Median	4,500	20	48	6.8

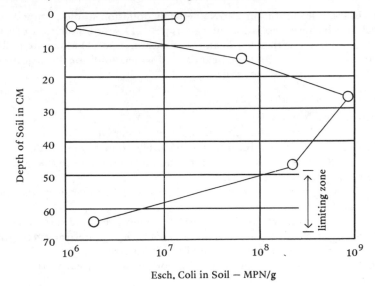

FIGURE 1. Concentration of coliform organisms with depth in Hanford soil.
(After Univ. of Calif., Tech. Bull. 13, pp. 43.)

The project is noted for its contribution to a knowledge of the behavior of both bacteria AND viruses with percolating water. Secondary sewage is given tertiary treatment in an oxidation pond and is then pumped to 6 percolation beds in parallel across a shallow stratum of sand and gravel confined in an old stream bed. The infiltered water moves down this confined channel a distance of 1,500 feet to an interceptor trench dug across the channel, passing enroute sampling wells located at 200 and 400 feet downstream from the percolation beds. Tracer studies showed the flow rate in the gravel to be about 200 feet in 2 days. Composite samples taken each day of a 10-day sampling period were analyzed for coliform bacteria, fecal coliform, and fecal streptococci with the results shown in Table 2.

The conclusions are that most removal occurred within the first 200 feet of travel, with little further removal in the next 1,300 feet above the trenches. Nevertheless, the data shows an impressive removal of bacteria by a *very coarse* medium.

II. Studies Outside the United States

Coliform bacteria labeled with radio phosphorus were injected in soils at the Tuzla Collective Farm in Rumania. Results showed that 92-97 percent of the coli were retained in the uppermost 1 cm of the soil tested, with 3-5 percent retained in the 1-5 cm layer.

The problem of ground-water pollution has been given intensive study in the Netherlands where many precautions are necessary because of the unique water situation; the water table is often less than 10 feet from the land surface.

J. K. Baars, a leader in current Netherlands pollution studies, studied (1957) chemical and bacterial pollution of two types: (1) severe pollution of surface layers of soil matter with small quantities of water; and (2) moderate pollution of surface layers of soil with large quantities of water. Additional categories were penetration into dry and wet soils.

Conditions and Results of the Studies

1. Penetration in dry soil.
   a. Test site located at a summer and winter campground.
   b. Each camp site was furnished with a privy 40 inches in depth.
   c. 10,000 kg (22 tons) of faecal matter and 400 M^3 (106,000 gallons) of urine was deposited each season.
   d. Drinking water was obtained from 3 wells 120 feet deep and located within the camp ground.
   e. Water levels ranged between 10-12 feet below the land surface.
   f. Soil-particle size was 0.17 mm (median), had a uniformity coefficient of 1.65, and extended to a depth of about 80 inches.
   g. Pollutants are washed below the bottom level of each privy by urine and rain water *only*.

Bacterial analyses of cores obtained directly adjacent to the tested privies revealed that *Eccherichia coli (E. coli)* could not be detected at depths exceeding 60 inches. Caldwell (1937 and 1938) attributes such a diminishing of coli to "soil defense," a mechanism by which soil bacteria, oxygenation, and nitrification kill most harmful bacteria before they travel an appreciable distance.

2. Penetration in wet soil.
   a. Large infiltration basins for towns and cities.
   b. Pollutant was tertiary treated sewage: bacteria count was 1,000-5,000 per ml; *E. coli* count was 100-200 per ml.
   c. The basin soils were generally comprised of sand of 0.15 mm median diameter.

Bacterial analyses of cores obtained next to the basins studied revealed a sharp increase in bacteria and *E. coli* counts but this is attributed to pre-existing conditions. *E. coli* was found as deep as 10 feet but rarely over 20 feet. Identical tests performed on virgin dune sand revealed that most harmful pollutants were removed in the first one foot and none were detected beyond 20 inches.

Baar concludes that:

1. Self purification requires time.
2. Harmful bacteria might travel 25 feet or more in very coarse material where the rate of ground-water flow is 25 feet or more per day.
3. A sand size of 0.15 mm or less is required.
4. Self purification occurs best in dry soils containing a sufficient supply of free oxygen.
5. In ground-water systems containing much pre-existing nutrient material self purification might require a much greater length of time.

6. Harmful bacteria is generally absorbed in the first 10 feet of travel due to oxygenation and nitrification.

Travel of Contaminants with Ground Water

Investigation of the movement of contaminants introduced directly into ground water in unconfined and confined aquifers has been reported by many workers. Miscellaneous studies and results presented in their order of occurrence follow.

Experiments in Germany (Ditthorn and Luerssen, 1909) involved the direct injection of bacteria-laden water in a well 177 feet deep. According to the authors the material penetrated below the water table consisted of sand and gravel in the following proportions.

Coarse gravel	-	6 percent	Coarse sand	-	28.8 percent
Medium gravel	-	5 percent	Medium sand	-	48.4 percent
Fine gravel	-	10.6 percent	Fine sand	-	1.2 percent

Reported porosity was 32.8 percent.

The first phase of the test took place during the summer. After injection of the pollutant the well was recharged at the rate of 380,000 gallons per day for 11 days.

At an observation well 60 feet distant bacteria were detected after nine days and the count ranged between 1 and 10 bacilli per 100 ml. The authors also reported that there was no increase in bacilli with time. Similar tests performed during the winter months revealed NO TRACE of bacteria in the observation well.

Stiles and Crohurst (1923) studied the pollution of ground water by privy wastes and excellently summarize their findings:

1. Pollution with fecal *Bacillus coli* has up-to-date been definitely and progressively followed in the ground water for distances of 3, 6, 10, 15, 25, 35, 45, 50, 55, 60, and 65 feet from the trench in which the pollution was placed; uranin has been recovered from these same wells and has spread to other wells at 70, 75, 80, 85, 90, 95, 100, 110, and 115 feet from the pollution trench. The soil in question is a fine sand with an effective size of 0.13 mm.
2. The pollution has traveled these distances within a period of 187 days, or about 27 weeks, and only in the direction of the flow of the ground water; no convincing evidence is present that the pollution has traveled against the flow of the ground water or at right angles to it.
3. The pollution has traveled only in a thin sheet at the surface of the zone of saturation; there is no evidence at present that it has dispersed radially downward, and even when heavy pollution is recovered at the top, water from lower levels (in near-by deeper wells) is negative both for uranin and for *B. coli*.
4. As the ground-water level falls, owing to dry weather, the pollution

tends to remain in the sand above the new (lower) ground-water level, namely, in the new capillary fringe.

5.   There is no evidence which would justify a conclusion at present that either the bacteria or the uranin is carried or moves to any appreciable distance in the capillary fringe itself, and there is neither theoretical reason nor experimental evidence to justify a conclusion that either the bacteria or the uranin progresses in the dry, aerated intermediate belt (between the capillary fringe and the upper soil belt). All present evidence is to the effect that when the ground-water level falls the pollution remains practically stranded in the capillary fringe or in the intermediate belt—according to the degree of fall of the ground water.

6.   A rainfall of 1 inch results in a rise of 5 to 6 inches in the ground-water table (in the particular experimental area in question); and if this rise is sufficient to re-establish the zone of saturation up at the level of the stranded pollution, the bacteria and the uranin are again picked up and carried along farther in the direction of the ground-water flow until dry weather again intervenes to cause another fall of the ground-water level.

7.   Thus the progressive (passive) movement and the stasis (stranding) of the pollution are intimately connected with, are dependent upon, and alternate with the rise and the fall of the ground-water level, and this latter factor is dependent upon the alternation of wet weather (rainfall) and dry weather (lack of rain at the intake area of the ground-water table). Experiments are now underway to determine, if possible, whether pollution placed directly into a deeper level of the ground water will travel up to the surface of the saturated zone.

8.   In explaining these results, capillarity, filtration, and gravity seem to come up for special consideration.

9.   In one experiment the pollution traveled only 45 feet from September, 1922, to May, 1923, and remained stranded at this distance. Study of the formation of the ground revealed that under the belt of pollution there is an impervious or nearly impervious stratum of peatlike material, which gradually tilts upward distally from the pollution pit and forms a ground-water dam; the pollution traveled out on high ground water to the dam, the ground-water level fell below the crest of the dam, and the pollution is now stranded, pending a rise of the ground-water table sufficient to produce a ground-water cascade which will carry the pollution over the crest of the dam.

10. The ultimate distance to which the pollution will be carried is dependent upon a number of complex and interlocking factors, namely, wet and dry weather, with resulting rise and fall of the ground water; the length of each of these periods; the rate of the ground-water flow (depending upon the "head" which, in turn, is dependent upon the rainfall); and, obviously, also the factor of the viability of the organisms under conditions of moisture, pH, food supply, etc., *ad finem.*

11. In another series of experiments human feces were buried in pits, in a locality of high ground water, and covered with sawdust. Of five samples

taken three years and two months after burial all were both macroscopically and microscopically recognizable as feces but the odor had become somewhat musty; three of these samples were positive and two were negative for *Bacillus coli;* ova of *Ascaris lumbricoides* were recognizable in all five samples, but all 57 ova found were dead.

12. The bearing of the foregoing results upon the intermittent pollution of wells, the location of water supplies, and the location of camps in peace or in war, will be evident to persons who are called upon for technical advice in these matters; the justification of the laws forbidding the use of abandoned wells for the disposal of excreta is self-evident; the possible effect of the custom (in some localities) of digging pits into ground water (as advised by some persons) is obvious.

13. In protecting wells, special attention should be given not only to surface protection as is now generally recognized but also to a new element, namely, the danger zone which exists from the highest water level to about a foot below the lowest water level. A leak in the pipe in this region is potentially very dangerous, and all wells unprotected in this danger zone are to be considered as potentially unsafe.

Caldwell (1937, 1938) intensively studied the nature of ground-water pollution caused by privies extending below the water-table level. Brief summaries of the conditions and results follow.

Bored Hole Latrine Study: Covington County, Alabama

1. Test site located in rolling upland beds of sand and sandy clays, some fine gravels.
2. Pollution stream was sustained by a clay stratum directly beneath the latrine boring.
3. The water-table gradient approximated 200 feet per mile. Depth to water approximated 11 feet.
4. Latrine was 15 inches in diameter, 16 feet deep, and penetrated approximately 5 feet into the ground water.
5. Fecal material from a family of six individuals was emptied into the latrine once daily.
6. Observation wells were located at 5-foot intervals up to 35 feet.
7. Observations extended to 379 days.

For the first few days bacterial pollution down-gradient from the latrine was not extensive. Most *B. coli* died before they were carried 5 feet at the rate of ground-water flow (about 7 feet per day). A few coli were detected at 15 feet in three days. After 5 weeks *B. coli* were detected at 10 feet. Chemical pollution extended beyond 35 feet.

After a period of two months *B. coli* were detected in 90 percent of the samples at 25 feet, and an occasional organism was detected at 35 feet.

After a period of 7 months soil "defense mechanisms" caused a retreat of the bacterial stream practically to the latrine.

Pit Latrine Study: Covington County, Alabama

1. Test site located on pastureland overlying beds of fine sand with some coarse sand and fine gravel.
2. Pollution stream sustained by a clay bed.
3. Water-table gradient approximated 35 feet per mile.
4. Depth to ground water ranged between 5 and 6½ feet. Velocity of flow was about 13 feet per day.
5. Latrine was 3 x 3 feet and 8 feet deep, and sheathed with 1 x 6 boards; ½-inch spaces left between lower 3 boards.
6. Fecal matter added as in bored hole latrine study.
7. Observation wells located at 10-foot intervals up to 80 feet.
8. Observations extended to 16 months.

Initially *B. coli* in significant numbers were carried 80 feet. At termination of the experiment "soil defense" had reduced the length of the *B. coli* streams to about 50 feet where only an occasional *B. coli* was recovered.

Pit Latrine Study: Covington County, Alabama

1. Test site located on a stream terrace 1,300 feet from any drainage system and 40 feet above the main river.
2. The 3 x 3 x 8½ feet deep latrine bottomed on permeable strata.
3. Water-table gradient was about 11 feet per mile.
4. Velocity of ground-water flow ranged between 2 and 3 inches per day.
5. Excreta from a family of 9 individuals were dumped daily into the latrine.
6. The study extended for a period of one year.

B. coli advanced slightly beyond 10 feet in 3 to 4 months when restriction of flow from the pit and soil clogging of the filtration bed caused regression. At termination the apex of *B. coli* barely reached 5 feet. Maximum lateral expansion of the stream was 5 feet after six months; rapid contraction followed. Maximum recovery of *B. coli* in 3 to 4 months approximated 200 per ml at 5 feet.

Anaerobes probably reached 50 feet in significant concentrations.

Similar studies were undertaken when the latrine was constructed such that the water-table level fluctuated between 12 and 13 feet below the bottom of the pit. Results of this study were:

1. From a dry pit latrine colon organisms did not penetrate 1 foot below or 1 foot laterally from the latrine.
2. When subjected to rains coli were carried up to but not exceeding 4 feet below the pit. Lateral spread was 1 foot.
3. When 100 gallons of water were added daily coli traveled up to but not exceeding 7 feet below the pit. Lateral spread was 2 feet.

The limitation of pollution was attributed to rapid "soil defense."

McGauhey and Krone (in SERL Report No. 67-11, 1967) studied pollution travel during direct recharge of a confined aquifer. Coliform bacteria and fecal streptococci were used as indicators. The major results indicated that in no instance did bacteria travel as far as 100 feet from the recharge well. Coliform and enterococci showed no difference in travel.

Additional Bacteria Studies

Other investigations conducted by the University of California reveal nothing in disagreement with studies described earlier. Experimental conditions of one particular project were somewhat different in that one of the studies involved injection of pollutants within a confined aquifer (Gotass, et al., 1954). Water containing 27 percent sewage was pumped into the aquifer for 38 days. The following facts were revealed.

The aquifer is located between 90 and 100 feet below the ground surface and consists of a three to seven-foot stratum of sand ranging in particle size from 0.21 mm to 0.71 mm. It is overlain and underlain by impervious clay. Measured transmissibility and permeability is 200 gpd/ft and 1,600-1,900 gpd/ft^2 respectively. Ground-water velocities ranged from 0.009 ft/min to 0.15 ft/min; ground-water movement was from north to south. After 38 days of continuous injection of pollutant at a rate of approximately 37 gallons per minute investigators found that maximum travel of coliforms was 100 feet south of and 63 feet north of the recharge well. The latter is interesting because it indicates that pollutants can migrate upstream relative to normal ground-water movement. The upstream migration was probably caused by the "mounding effect" of ground-water recharge. Travel time for the maximum migration was 33 hours.

The California investigators have developed a formula for computing the reduction of the coli count per foot of travel but they point out that the formula can be applied ONLY after the aquifer had been subjected to a series of tests to determine values for various constants which are incorporated in the formula. Calculations for the aquifer tested show that bacteria reduction for 20-30 percent sewage averages about 25 percent reduction per foot of travel.

Studies involving the viability of bacteria in porous media report that under favorable living conditions some forms of bacteria, including the types mentioned in this report, may survive up to five years. It should be mentioned that such conditions are extreme (tropics) and that 60 to 100 days might be the maximum life in temperate climates. Most investigators agree that the nature of the soil in contact with the source of contamination plays a dominant role in the subsequent life and travel of bacteria.

Virus Studies

Studies of the travel of viruses with percolating waters reveal that some types behave like bacteria as far as initial die off is concerned but that certain types which are more resistant to "environmental changes" might travel far greater distances than their bacterial counterpart. Although there are many reports of viruses traveling more than 50 feet and causing various sicknesses and epidemics, most controlled experiments with viruses indicate a tendency to die within 10 feet of their source. It is obvious that more detailed studies with virus contamination are needed.

Because of the potential dangers existent in the study of virus movement in porous media most such experiments are conducted under very rigidly controlled conditions. This is somewhat unfortunate because it has prevented us from acquiring a knowledge of how viruses are influenced by NATURALLY EXISTING CONDITIONS. Available reports on the subject indicate that the viruses studied are influenced by the same factors as are bacteria. That is, their movement in porous media is governed by filterability. Because of the very small size of viruses (relative to the size of bacteria) the factors which make up a media's filterability are somewhat restricted. Several investigations have indicated that virus removal is primarily a result of adsorption and die off.

Studies performed for the Department of the Army have indicated that for sands ranging in classification from fine-clayey sand to coarse-granitic alluvium certain viruses are more effectively retained in the finer sands, particularly if they contain a relatively high percentage of clay and silt. Virus removal was shown to increase with decreasing particle size. The greater percentage of removal took place in the uppermost portion of the sand columns tested. It was shown that for a well-sorted sand of particle size averaging 0.12 mm the removal efficiency for two feet of penetration was 99.999 percent. The report emphasized that workers in the field should be careful not to make false interpretations in evaluating sand-filter removal capabilities. This is particularly important in the evaluation of many bacterial and virus studies where the investigator might be blinded by removal capability figures and tend to disregard data which indicate the INITIAL CONCENTRATION of pollutant. Inclination: It is possible that higher concentrations of viruses would result in proportionately longer lengths of travel.

The viruses studied for the Department of the Army include the bacteriophages T-1, T-2, and phage 65. These agents are not TRUE viruses but are merely virus-like in regard to both size (about 0.5 micron) and capability of causing diseases of animals and plants.

The University of California Sanitary Engineering Research Laboratory at San Diego conducted a test wherein both bacteria AND viruses were subjected to pollution studies (the Santee Project). The results of the bacteria phase was described earlier. In a virus study 12 liters (about three U.S. gallons) of concentrated polio vaccine virus type 3, was mixed with 10 gallons of unchlorinated water and applied to the percolation bed influent over a period of three hours. Virus samples were then collected from the previously described sampling wells at 200 and 400 feet downstream and from the infiltration trench at 1,500 feet. All samples were found to be negative and it was concluded that the virus had been removed in less than 200 feet of travel. Recall that the project was located within the confines of an old stream bed. Channel fill consisted of fine sand through coarse sandy gravel (specific particle size data not available at the time of this writing).

Over a period of three years the infiltration ponds were subjected to raw sewage, activated sludge effluent, and oxidation pond effluent. Reovirus,

Adenovirus, and several types of polio, ECHO, and Coxsackic viruses were at times identified in the oxidation pond effluent; but never found in the percolate from the soil system. It is interesting to note (and indeed unfortunate) that observation wells were not made available near the oxidation ponds and for the length of stream between the percolation field and the first trench 200 feet downstream. All that can be concluded from this experiment is that the viruses did not migrate as far as 200 feet from the source of contamination.

Experiments with Polio Type 1 virus were performed by G. G. Robeck (in McGauhey and Krone, 1967) who reported a complete and continuous removal of the virus within a two-foot depth of sand. Various sands were subjected to the treatment; all had good filter capabilities.

Workers at Colorado State University studied the movement of virus-sized particles (albumin molecules: 0.015 micron) and reported up to 96 percent retention for travel through less than three feet of Greeley sand (an extremely sandy agricultural soil) previously heated to 1,400 degrees F. in a muffle to destroy organic matter and any microorganisms which might have been present.

Summary

A list of outstanding characteristics of the movement of biological pollutants through porous media can now be presented.

1.   Bacteria and viruses travel with the flow of water; they do not travel or move against the current.

2.   Bacteria and viruses can move in a direction opposite to that of normal ground-water gradients during times of recharge and/or pumping. The latter is an assumption based on the characteristics of well hydraulics.

3.   In general bacteria and viruses are removed by the aquifer in the same manner as are coliforms.

4.   The rate of bacterial and virus removal with distance is a function of an aquifer characteristic termed "filterability."

5.   For any degree of filterability it depends upon distance only and NOT UPON THE RATE OF POLLUTANT RECHARGE.

6.   Aquifer materials best suited for the removal of biological contaminants are those uniformly composed of very fine to fine grained sand with a high clay content.

7.   For an ideal system the maximum length of travel of biological pollutants with ground water ranges between 50 and 100 feet.

8.   Pollution travel in nonsaturated systems is considerably less than that in saturated systems in that maximum lengths of travel appear to be in the vicinity of 10 feet.

9.   The nature of the soil in contact with the source of contamination plays a dominant role in the subsequent travel of bacteria.

10. Bacteria and/or virus-infested pollutants might travel much farther than predicted if nutrient-laden waters are intercepted during the course of penetration.

11. Under favorable conditions (not necessarily applicable to Colorado) bacteria and viruses have been known to survive up to five years.

Because of the many variables involved in the determination of a "safe" distance of a domestic well from a potential source of pollution no single set of distances will be reasonable for all conditions. Because of the diverse geologic conditions which prevail throughout the State it is suggested that domestic well applications in areas of existing or potential ground-water contamination be given the following considerations:

I.   Plumbing systems within the "area of influence."
    1.   Standards for sewer-pipe junctions.
    2.   Septic tank construction standards.
    3.   Absorption system standards.
    4.   Disinfectant facilities.
    5.   Character and location of existing and potential pollution sources.
II. Hydrology and geology of the "area of influence."
    1.   Direction of surface runoff.
    2.   Flooding considerations.
    3.   Bedrock configuration.
    4.   Texture of material between source of contamination and point of juncture between well casing and aquifer.
    5.   Depth to water with respect to source of contamination.
    6.   Hydraulic properties of the aquifer.
    7.   Probable direction of movement of contaminants with respect to the well during times of pumping and during the possible creation of a ground-water mound during periods of high discharge of pollutant-laden water from a point or line source.
    8.   Other.
III. Well statistics.
    Consideration of all standards necessary to prevent contamination from existing and potential sources.
IV. Close scrutiny of engineering report forms.
V.   Annual monitoring of all wells within potential or existing problem areas.

All of these considerations might not be feasible for any given situation. They are intended to give the reader an idea of the existing and potential complexities involved in sound pollution-safe distance determinations.

A summary of all the material examined for this report indicates that each individual stratum of porous media possesses its own characteristic filterability. Since each individual stratum possesses its own filterability the use of a single-line graph indicating the removal of bacteria and viruses with distance would be somewhat hazardous. In place of a single-line graph, double-line graphs indicating the approximate minimum and maximum travel for biological pollutants are presented. The graphs are based on well-documented case histories and could be used only in conjunction with the previously mentioned considerations.

Three "zones" are illustrated on the graphs. The "prohibitive" zone indicates those regions of particle size with respect to distance from a source of pollution that a domestic well should not be constructed. The "hazardous" zone envelops the region for which examined case histories are well documented. It is meant to illustrate the zone or distance biological pollutants are known to have traveled in a medium of given particle size. It is important to note that the lowermost line of the hazardous zones is based on maximum lengths of travel. The "probably safe" zone represents those regions of particle size with respect to distance from a source of pollution that a domestic well can probably be constructed in reasonable safety.

Two graphs are presented (Figures 2 and 3)—one for pollution travel in areas of nonsaturation and one for pollution travel with the flow of ground water. Both graphs were constructed on the basis of studies in sedimentary materials. Travel of pollutants in "hard rock areas" has not been studied but it can probably be safely assumed that pollutants intercepting interconnecting rock fractures or open cavities and the like might travel for a considerable distance.

The problem of determining "safe distances" between domestic wells and pollution sources is somewhat difficult from the human standpoint because rules and regulations for water well drillers and pump installation contractors must be designed to safeguard public health and at the same time must be reasonably fair to the landowner who wishes to have his own source of domestic water and to the licensed water well driller whose livelihood need no explanation.

Table 3 illustrates the comparison between distance recommendations by the U.S. Public Health Service, the Federal Housing Authority, and the States of California and Colorado. Other States generally follow suit with the FHA or USPHS.

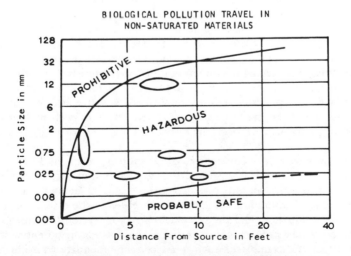

FIGURE 2. Biological pollution travel in nonsaturated materials.

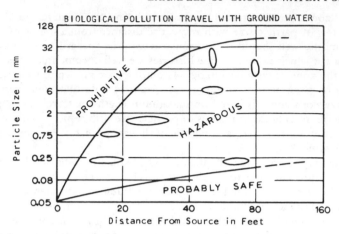

FIGURE 3. Biological pollution travel with ground water. Zones of maximum length
of travel of biological pollutants as described in documented case histories.

Recommendations

Regulations regarding spacing requirements between domestic food processing
wells and potential or existing sources of contamination should be made more
flexible than the present standards. Spacing requirements should, whenever
possible, be based on local factors because no one set of distances will be
adequate and reasonable for all conditions. In areas where "undesirable"
conditions exist safe distances should be extensive. In areas possessing especially
"favorable" conditions lesser distances might be acceptable if approved by the
enforcing agency. In the determination of the factors which will govern the
selection of a safe distance the following MINIMUM factors should be evaluated:
(1) the character and location of the sources of potential or existing
contamination; (2) geologic and hydraulic characteristics of the material
between the land surface and water table, the water table and bedrock, or
whatever might be applicable; (3) seasonal depth to water, its direction and rate
of movement, and the mathematically derived effect of well pumping on the
direction of ground-water movement between the source of contamination and
the well.

In lieu of extensive examination of a subject area's hydrogeologic and
sewerage characteristics, it is suggested that the method of determination of safe
distances be broken down into two categories—the first being those systems in
which the zone of saturation lies some distance *below* the existing or potential
source of contamination, and the second being those systems in which the zone
of saturation is *very near* or *in contact with* the existing or potential source of
contamination. An example of such a set of recommendations is illustrated in
Table 4. Schematic diagrams of typical situations are presented in the appendix.

Table 4 is designed primarily for areas underlain by unconsolidated

TABLE 3. Comparison of recommended safe distances between domestic wells and pollution sources as specified by the FHA, the USPHS, and the States of California and Colorado.

| Source of Pollution | Recommended Distance in Feet | | | |
	California	Colorado	FHA	USPHS
Septic Tank	50	(1) 100 feet from point of juncture between well casings and aquifer.	50	50
Sewer Lines with Watertight Joints	10		10	10
Other	50		50	50
Percolation Field	100[1]	(2) Minimum horizontal distance between well casings and potential source of contamination is 25 feet.	100[1]	100[1]
Absorption Bed	100[1]		100[1]	–
Seepage Pit	100[1]		100[1]	100
Drywell	50		50	50
Cesspool	–		–	150
Other	as recommended		as recommended	–

[1] Distance (horizontal) may be reduced to 50 feet if point of beginning of well casing perforations and percolating field are separated by a well-defined continuous impervious strata.

TABLE 4. Example of recommended safe distances between domestic wells and various sources of pollution.

Source of pollution	Areas in which the zone of saturation is greater than 10 feet from the source of contamination	Areas in which the zone of saturation is less than 10 feet from the source of contamination
Septic tank with absorption trench or seepage bed	25-100[1,2,3]	75-100[1,2]
Septic tank with seepage pit. (If possible pit should terminate at least four feet above water-table level)	50-100[1,2]	75-100[1,2]
Septic tank with no absorption system	do	do
Sewer lines with water-tight joints	10[2]	25[2,3]
Abandoned well	25[2,3]	as recommended
Cesspool (not recommended)	50[2]	as recommended
Other	as recommended	as recommended

[1] The distance is to be measured from the well to either the septic tank outlet pipe or a particular edge of the absorption system whichever is closer to the well.
[2] Distance should be reduced or extended depending upon the individual situation or as recommended by the enforcing agency.
[3] For spacing in the vicinity of 25 feet, wells should be properly sealed to a depth of no less than 30 feet.

sedimentary material such as those occurring in the plains east of the Rocky Mountains and mountain valleys whose floors are composed of alluvium. Structural basins such as North Park, South Park and the San Luis Valley are included.

Areas underlain by igneous, metamorphic, or consolidated sedimentary rocks should be given a more critical examination because of the danger of extensive pollution travel via fractures and solution cavities. In lieu of an extensive survey the minimum distance between source of pollution and the well should be 100 feet unless it can be shown that a shorter distance will not increase the probabilities of contamination.

Annual monitoring of wells is recommended for areas having a relatively dense population of both domestic wells and soil absorption systems. The enforcing agency should also consider informing the public on associated hazards.

Definitions

Absorption Trench. A trench 12 inches to 36 inches in width which contains aggregate and distribution pipe, and accepts effluent from a septic tank.

Seepage Bed. A trench exceeding 36 inches in width. Its function parallels that of an absorption trench.

Seepage Pit. A covered pit with lining designed to permit effluent from a septic tank to seep into the surrounding soil. The USPHS does not recommend the use of seepage pits if there is any likelihood of contaminating underground waters. When seepage pits must be used in lieu of a better system, the pit excavation should terminate 4 or more feet above the water table.

Cesspool. An excavation designed to accept both liquid and solid wastes from a drainage system. Cesspools are designed to retain the organic matter and solids, but permit seepage of liquids through their bottom and sides. They are NOT RECOMMENDED as a means of sewage disposal. Cesspools should be constructed as far as possible from sources of water supply.

Definition of Various Terms

Aerobic, anaerobic, anaerobe. Aerobic is a condition wherein molecular oxygen is present. A condition lacking molecular oxygen is classed as anaerobic. An anaerobe is a microorganism capable of surviving in anaerobic conditions, etc.

Coliform bacteria. A microorganism common in the intestinal tracts of man and warm-blooded animals. Coliform bacteria are commonly used as indicators of the possible presence of harmful bacteria because where they are found it is assumed that typhoid, dysentery and other harmful bacteria from the intestinal tract may be present.

Eccherichia coli (Escherichia coli, E. coli). A type of coliform bacteria which may infect the urinary tract of man and cause cystitis.

Fecal Bacillus coli, fecal coliform. General terms for those bacteria which have their natural habitats within the intestinal tract of man and beast.

Fecal streptococci (Streptococcus fecalis). An *a*-hemolytic bacteria which brings about the dissolution of the red-blood cells of higher animals. Enterococci is a general term.

M P N. Most Probable Number. A figure expressing a statistical "count" of the number or quantity of microorganisms over a given area or within a given volume.

Nitrification and Oxygenation. Complex physiochemical processes by which various bacteria within a soil-water system "fix" nitrogen and oxygen as components of various insoluble compounds which might aid in the process of soil clogging. Many investigators consider the terms misnomers.

Pathogenic organism. An organism capable of causing disease.

Virus. An agent of disease which can be transmitted indefinitely in series, which cannot be seen with a standard optical microscope, cannot be filtered or separated by ordinary methods, and cannot be cultivated on lifeless mediums which support the growth of bacteria. Viruses are smaller than bacteria and range in size from about 0.08 micron to about 0.7 micron; bacteria range in size from about 0.2 micron to about 15 microns.

Appendix

Schematic diagrams of conditions which might influence the determination of the "safe distance" between a well producing water for human consumption and/or food processing and a source of potential or existing contamination. The diagrams are not designed to represent all possible conditions; they merely illustrate several of the variables which should be considered in the process of determining a reasonably "safe distance" between a well and a point or line source of contamination.

The following symbols are used:

⬛⬛⬛   Septic tank with absorption trench or seepage bed

⬜⬛⬛⬛   Septic tank with seepage pit

⬛⬛⬛   Septic tank with no absorption systems

[C]   Cesspool

◯   Sewer line

‖   Abandoned well

All diagrams cut sections of unconsolidated sand and gravel unless otherwise indicated.

SEPTIC TANK

Areas in which the zone of saturation is greater than
10 feet from the source of contamination.

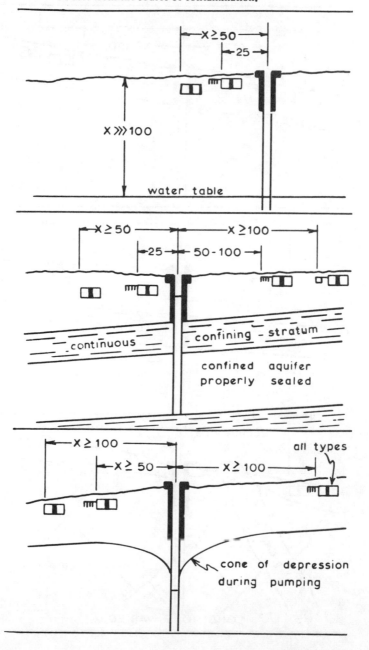

SEPTIC TANKS

Areas in which the zone of saturation is less than
10 feet from the source of contamination

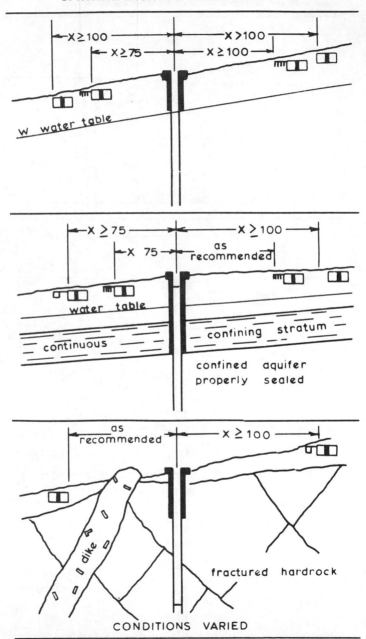

Areas in which the zone of saturation is greater
than 10 feet from the source of contamination

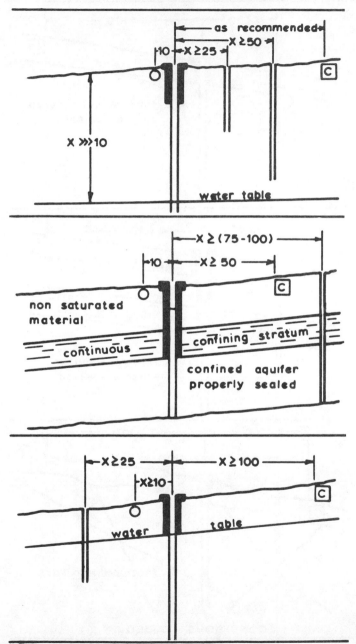

Areas in which the zone of saturation is less than
10 feet from the source of contamination

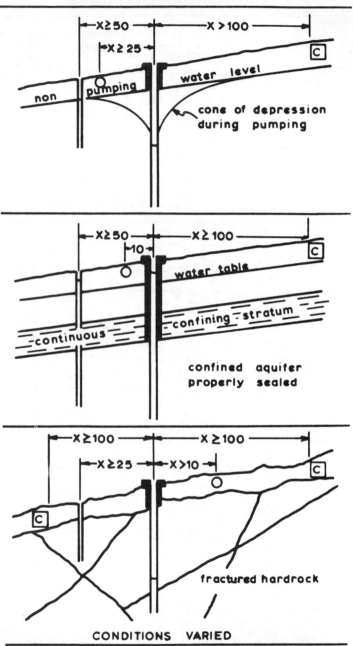

CONDITIONS VARIED

References

1. Baars, J. K. 1957. Travel of pollution and purification enroute in sandy soils. Bull. World Health Organ. 16 (4): 727-747.
2. Bailey, G. W. 1968. Role of soils and sediment in water pollution control. U.S. Dept. Interior, Federal Water Pollution Control Commission, Part 1.
3. Burrows, W., et al. 1959. Textbook of microbiology. Philadelphia and London: W. B. Saunders Company.
4. Butler, R. G., G. T. Orlob, and P. H. McGauhey. 1954. Underground movement of bacterial and chemical pollutants. J. Amer. Water Works Assoc. 46:97-111.
5. Caldwell, E. L. 1937a. Study of an envelope pit privy. J. Infectious Disease 61; 264-269.
6. Caldwell, E. L. 1937. Pollution flow from pit latrines when an impervious stratum closely underlies the flow. Infectious Disease 61; 270-288.
7. Caldwell, E. L., and L. W. Parr. 1937. Ground-water pollution and the bored-hole latrine. J. Infectious Disease 61; 148-182.
8. Caldwell, E. L. 1938a. Pollution flow from a pit latrine when permeable soils of considerable depth exist below the pit. J. Infectious Disease 62; 219-258.
9. Caldwell, E. L. 1938b. Studies of subsoil pollution in relation to possible contamination of ground water from human excreta deposited in experimental latrines. J. Infectious Disease 62:271-292.
10. Calvert, C. K. 1932. Contamination of ground water by impounded garbage waste. J. Amer. Water Works Assoc. 24: 266-276.
11. Clarke, N. A., and S. L. Chang. 1959. Enteric viruses in water. J. Amer. Water Works Assoc. 51: 1299-1317.
12. Dappert, A. F. 1932. Tracing the travel and changes in composition of underground pollution. Water Works and Sewerage 79 (8): 265-274.
13. Ditthorn, F., and A. Luerssen. 1909. Experiments on the passage of bacteria through soil. Engr. Record 60; 642.
14. Drewry, W. A., and R. Eliassen. 1968. Virus movement in ground water. J. Water Pollution Control Federation, Part 2, p. R 257-R 271.
15. Eliassen, R., P. Kruger, W. Drewry, G. Tchobanglous, et al. 1964—1967. Studies on the movement of viruses in ground water (four annual progress reports). Water Quality Control Research Lab., Stanford Univ. Stanford, California.
16. Filmer, R. W., and A. T. Corey. 1966. Transport and retention of virus-sized particles in porous media. Colorado State University, Ft. Collins, Colorado, Sanitary Engineering Paper 1.
17. Fournelle, H. J. 1957. Experimental ground-water pollution at Anchorage, Alaska Amer. J. of Pub. Health 72:208-209.
18. Gotaas, H. B., et al. 1954. Report on the investigation of travel of pollution. California State Water Pollution Control Board, Sacramento, Calif., Pub. 11.
19. Mallmann, W. L., and W. Litsky. 1951. Survival of selected enteric organisms in various types of soil. Am. J. Pub. Health 41:38-44.
20. McGauhey, P. H., and R.B. Krone. 1967. Soil mantle as a wastewater treatment system. Sanit. Engr. Research Lab., Univ. of Calif., S.E.R.L. Rept. 67-11.
21. McLean, D. M. 1964. Contamination of water by viruses. J. Amer. Water Works Assoc. 56:585-591.
22. Orlob, G. T., and R. B. Krone. 1956. Movement of coliform bacteria through porous media. Sanit. Engr. Research Lab., Univ. of Calif., Final Rept., U.S. Public Health Service Grant 4286.
23. State of California. 1968. Water well standards. California Dept. Water Resources Bull. 74.
24. University of California, 1955. Studies in water reclamation. Sanit. Engr. Research Lab. Univ. of Calif., Tech. Bull. 13.
25. U.S. Dept. Health, Education, and Welfare. 1961. Ground-water contamination. Proceedings of the 1961 symposium on ground-water contamination, Robert A. Taft Sanitary Engineering Center, Cincinnati, Ohio.
26. Warrick, L. F., and E. J. Tully. 1931. Pollution of abandoned well causes Fond du Lac typhoid epidemic. Engr. News Record 104:410-411.

Part Five

Trace Elements—
A New Factor In
Water Pollution

S everal types of potentially toxic chemicals such as certain trace elements are present in both treated and raw drinking water supplies. These substances may be the result of overt industrial pollution, easily traceable to the source, or they may originate from more subtle causes such as the solution of the metal in pipes and pumping plants used to transmit the treated water to the consumer. In other instances trace elements occur in water from natural causes.

The chemistry of natural water is enormously complex and is far from being well understood. Some 12,000 different toxic chemical compounds are involved in various industrial processes today and scores of new types are formulated each year. Most of these compounds find their way into wastewater. Unfortunately, little is known of the environmental impact of many of these chemicals— especially the effect of prolonged low-level dosages on human health. Moreover, the problem is compounded because a determination of the significance of industrial contaminants is extremely complicated. Pollutants, whose specific identities may be largely unknown, may enter the water as a complex mixture of many substances of diverse chemical natures. Positive identification of a specific contaminant may be exceedingly difficult.

The development within the last decade of sophisticated analytical equipment

Photo opposite courtesy of *Minneapolis Tribune,* Minneapolis, Minn.

and techniques has opened a nearly virgin field of investigation, allowing researchers to detect extremely small concentrations of chemical substances in food, water, and air. Although these materials can now be detected, little is actually known about their composition, distribution, and physiological effects. This makes the task of establishing realistic limits in consumer products very difficult.

Another major complicating factor is the chemical occurrence of certain contaminants. In one chemical state a trace element may pass through the body with little or no harm, but in another it may be absorbed to the point of toxicity. An example of this phenomenon is the difference in health effects between elemental mercury and methyl mercury.

The report by W. H. Durum and others indicates that of the seven elements they looked for in surface water throughout the nation, all of them appeared in only minute concentrations and only in a few instances did the concentrations exceed the recommended Public Health Service limit. The reports by H. V. Warren and R. E. Delavault and by W. A. Pettyjohn, however, point out that even though certain heavy metals occur in very small amounts in water, these same substances may become highly concentrated in soil and river mud, removed by plants and animals, and thus enter the food chain in increasing concentration.

Reconnaissance of Selected Minor Elements in Surface Waters of the United States, October, 1970

W. H. Durum*, J. D. Hem**, and S. G. Heidel***

Summary

A nationwide reconnaissance of selected minor elements in water resources of the 50 States and Puerto Rico was made by the U.S. Geological Survey in cooperation with the U.S. Bureau of Sport Fisheries and Wildlife during autumn, 1970. Initiated in response to the growing need for data on minor elements in water, including toxic metals, the synoptic survey provides an up-to-date baseline of such data largely for dry-weather flows of streams that are sources of municipal and industrial water for the Nation's metropolitan areas.

More than 720 samples obtained during October and November 1970 from rivers and lakes were analyzed for arsenic, cadmium, chromium (hexavalent), cobalt, lead, mercury, and zinc. Sampling sites fall within three categories: (1) surface water sources of public water supplies for cities of more than 100,000 population, or for some states, the largest city in each State, (2) water courses downstream from major municipal and (or) industrial complexes in each State, and (3) U.S. Geological Survey hydrologic bench-mark stations. Hydrologic bench-mark stations were established in the mid-1950's for measuring long-term natural trends in streamflow and water quality. These stations are located in undeveloped drainage basins in the major physiographic regions of the country.

Samples were taken mainly during the period October 1-15, 1970, when rivers were at medium or low-flow stages in many parts of the country. Samples were filtered (0.45-micron openings) to provide information on sediment-free water like that normally supplied to the user. The samples were acidified (1.5 milliliters nitric acid for a l-liter sample) when collected to prevent separation of minor elements during shipment to a designated laboratory for analysis. For each site, stream discharge in cubic feet per second was measured or estimated at the time of sampling, when it was practicable to do so. More complete chemical

Reprinted from U.S. Geological Survey Circular 643, 1971, pp. 1-7. Analytical data (table 1) have been deleted.

*Water Resources Division, U.S. Geological Survey, Washington, D.C.
**Water Resources Division, U.S. Geological Survey, Menlo Park, California.
***Water Resources Division, U.S. Geological Survey, Parkville, Maryland.

analyses for samples collected at many of these sites have been published in the Geological Survey water resources data reports.

The sampling sites used in this survey are shown in figures 1—6 in the back of the report. A brief summary of the results for each element is given below.

Arsenic. Seventy-nine percent of the 727 samples examined had arsenic concentration less than 10 μg/1 (micrograms per liter) which is the lower limit of detection, 21 percent of the samples had arsenic greater than 10μg/1, and 2 percent had more than 50μg/1—the maximum considered safe for drinking water (U.S. Public Health Service, 1962). A sample from Sugar Creek near Fort Mill, S.C., downstream from an industrial complex in North Carolina, had the highest arsenic content, 1,100 μg/1, of any sample obtained in this survey.

Arsenic was identified in as many samples from bench-mark locations as in samples from any other sources.

Cadmium. Cadmium was detected in 42 percent of the samples in concentrations ranging from 1 to 10 μg/1. About 4 percent of the river samples had cadmium in excess of 10 μg/1 (the Public Health Service upper limit for drinking water) and these occurred in about one-third of the States. The maximum concentration found was 130 μg/1. Cadmium was detected in samples from bench-mark sites, public water supplies, and metropolitan-industrial complex locations; but 54 percent of the samples did not contain measurable amounts of cadmium. The implication of the data is that the highest concentrations of cadmium in water generally occur in areas of high population density.

Chromium. Chromium (hexavalent) rarely was detected at levels much above about 5 μg/1 and occurred in the range 6 to 50 μg/1 in only 11 of more than 700 samples. There were no concentrations in excess of 50 μg/1, the upper limit for hexavalent chromium in drinking water.

Cobalt. Cobalt concentrations most commonly were below the detection limit (less than 1 μg/1), but cobalt was found in 37 percent of the samples, commonly in the range from 1 to 5 μg/1. The higher value is about the upper limit of solubility of cobalt in normal river water.

Lead. Lead was found in about 63 percent of samples in concentrations ranging from 1 to 50 μg/1. In a few waters lead was detected in excess of 50 μg/1, the upper limit for drinking water. Lead occurs widely in the range 6 to 50 μg/1.

Lead was detected less frequently in samples collected at bench-mark sites than in those from public water-supply sources and from streams below metropolitan-industrial areas.

Mercury. Data on mercury are reported in two forms: dissolved and total. The concentration of the dissolved form is an indication of what might occur in a treated or filtered water supply at the same sampling point. Total mercury represents the amount in the water-sediment mixture. The difference between dissolved and total mercury is indicative of the portion that adheres to suspended particles which might be a part of the food chain of the aquatic community.

Dissolved mercury ranged from below the lower limit of detection (0.1 μg/1) to 4.3 μg/1 and was found in only 7 percent of the samples. In none did the concentration exceed the proposed Public Health Service upper limit for dissolved mercury in drinking water, which is 5 μg/1. Total mercury was found in excess of 5 μg/1 in a few instances.

Zinc. The concentration of zinc in most samples was in the range of 10 to 50 μg/1, but occasionally exceeded 5,000 μg/1, the recommended (not mandatory) upper limit for drinking water.

The survey shows, as one would expect, that the heavy metals studied are widely distributed in low concentrations in water. There is some evidence that the concentration levels are related to man's activities in certain instances. There appears to be no widespread occurrence of these metals in water in amounts exceeding current drinking water standards. However, the initial assessment of these data does indicate potenital problems in a few areas. Although firm conclusions regarding natural patterns and pollution anomalies cannot be drawn from the first results of a survey of this kind, similar data collected in the future will show whether trends exist in the overall distribution of these metals in water and whether the observed anomalies persist and should be studied in greater detail.

Units and Terms

Throughout the report, the following units have been used:

μg/1 (*microgram per liter*). Equivalent to 1 part per billion (ppb) or 1 pound in a billion pounds of water.

Cfs (*cubic foot per second*). 1 cfs is equivalent to 0.65 mgd (million gallons per day) or 0.0283 cubic meter per second.

Duration. Percentage of days (on an annual basis) in which the flow equals or exceeds that given.

Summary of Analytical Methods

Geological Survey analytical methods (Brown and others, 1970) were used uniformly by all participating laboratories. A provisional Federal Water Quality Administration method (written commun., 1970) for total and dissolved mercury was used.

1. *Arsenic* (*silver diethyldithiocarbamate method*). Inorganic arsenic compounds are reduced to arsine by zinc in an acid medium. The resulting mixture of gases is passed through a scrubber containing pyrex wool impregnated with lead acetate solution and into an absorbing tube containing silver diethyidithiocarbamate dissolved in pyridine. Arsine reacts with silver diethyldithiocarbamate to form a soluble red substance having maximum absorbance at about 535 mμ. The absorbance of the solution is measured spectrophotometrically and arsenic is determined by reference to an analytical curve prepared from standards.

2. *Cadmium* (*atomic absorption method—chelation-extraction*). Cadmium in the sample is chelated with ammonium pyrrolidine dithiocarbamate (APDC). The

chelate is then extracted from the sample with methyl isobutyl ketone (MIBK), which is aspirated into the flame of a spectrophotometer.

3. *Hexavalent chromium (atomic absorption method—chelation-extraction).* Hexavalent chromium in water is chelated with ammonium pyrrolidine dithiocarbamate (APDC) and is extracted from the sample with methyl isobutyl ketone (MIBK) at a pH of 2.4. The MIBK is aspirated into the flame of the spectrophotometer.

4. *Cobalt (atomic absorption method—chelation-extraction).* Cobalt in water is determined by chelation with ammonium pyrrolidine dithiocarbamate (APDC). The chelate is then extracted from the sample with methyl isobutyl ketone (MIBK), which is aspirated into the flame of the spectrophotometer.

5. *Lead (atomic absorption method—chelation-extraction).* Lead in the sample is chelated with ammonium pyrrolidine dithiocarbamate (APDC), which is then extracted into methyl isobutyl ketone (MIBK) and aspirated into the flame of a spectrophotometer.

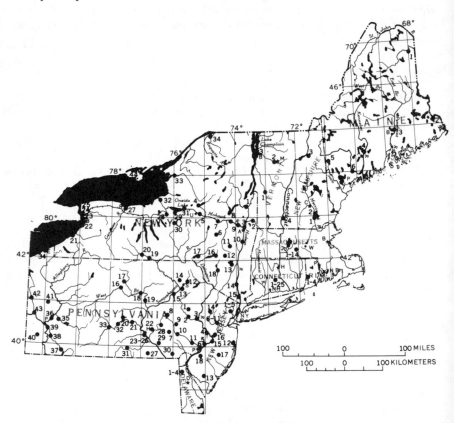

FIGURE 1. Map of the northeastern United States showing sites from which samples of minor elements were obtained, October 1970. Numbers correspond with those shown in table 1 [deleted].

FIGURE 2. Map of the southeastern United States showing sites from which samples of minor elements were obtained, October 1970. Numbers correspond with those shown in table 1 [deleted].

6. *Zinc (atomic absorption method—direct).* The sample is aspirated directly with no pretreatment other than dilution or filtration as may be required. Zinc concentrations are then determined by standard atomic absorption measurements.

7. *Mercury (flameless atomic absorption procedure).* Persulfate oxidation follows the addition of permanganate to insure that organo-mercury com-

FIGURE 3. Map of the central United States showing sites from which samples of minor elements were obtained, October 1970. Numbers correspond with those shown in table 1 [deleted] .

pounds, if present, will be oxidized to the mercuric ion before measurement in the spectrophotometer. The procedure determines total mercury in water or water-sediment mixtures without differentiating inorganic from organic. The procedure was used also for dissolved mercury (Federal Water Quality Administration, written commun., 1970).

8. *Mercury (silver wire atomic procedure).* Mercury is collected from an acidified sample of filtered water by amalgamation on a silver wire. The silver wire is electrically heated and the vapor drawn through an absorption cell placed in the light beam of the atomic-absorption spectrophotometer. The procedure was used in the study in some instances for determining with high precision (0.1 μg/1) that fraction of the total mercury present as stable inorganic mercury.

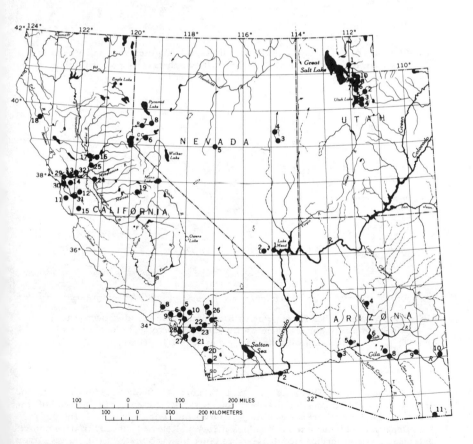

FIGURE 4. Map of the southwestern United States showing sites from which samples of minor elements were obtained, October 1970. Numbers correspond with those shown in table 1 [deleted].

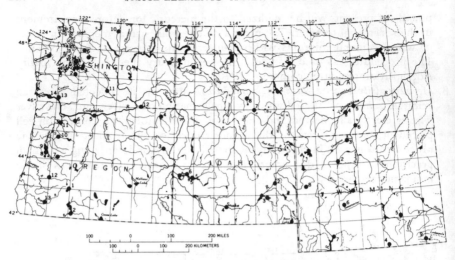

FIGURE 5. Map of the northwestern United States showing sites from which samples of minor elements were obtained, October 1970. Numbers correspond with those shown in table 1 [deleted].

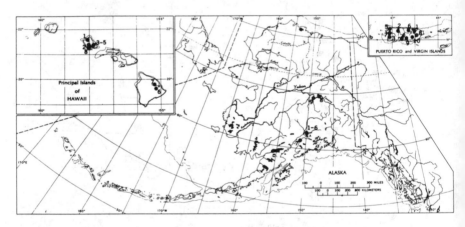

FIGURE 6. Map of Alaska, Hawaii, Puerto Rico, and Virgin Islands showing sites from which samples of minor elements were obtained, October 1970. Numbers correspond with those shown in table 1 [deleted].

References

1. Brown, Eugene, M. W. Skougstad, and M. J. Fishman. 1970. Methods for collection and analysis of water samples for dissolved minerals and gases. U.S. Department of the Interior, Geol. Survey Techniques Water Resources Inv., book 5, chap. A1, 160 p.
2. U.S. Department of the Interior, Geological Survey. Released annually. Water resources data, part 2, Water quality records (separate book for each State). Geol. Survey, Water Resources Division.
3. U.S. Department of Health, Education, and Welfare, Public Health Service. 1962. Drinking water standards. Public Health Service Pub. 956. 61 p.

A Geologist Looks at Pollution: Mineral Variety

Harry V. Warren* and Robert E. Delavault**

Fashion tends to change whether it be in men's beards, women's hats, or in the attitudes of society towards pollution. At the present time it seems to be quite fashionable to be pollution-conscious. Unfortunately, the subject of pollution is apt to generate in the minds of some people a disposition to approach the subject on a basis of emotion rather than of cold logic.

Before further discussion, let us attempt to define what is meant by "pollution." Doubtless there are many definitions, but the two that follow should provide adequate guidelines for the few remarks I wish to make. Both were provided by speakers at the Resource Ministers Conference on Pollution. The first was by John P. Tully:

> Pollution is taken to mean alteration of the natural environments, air, water, and soil, so that they are rendered offensive or deleterious to men's aesthetic senses or uses, or to animals, fishes, crops which man wishes to preserve. It is recognized that some degree of alteration of the environments is a necessary consequence of man's activities. Such alterations are not considered pollution until they reach the limit of tolerance(1).

Because much of our direct involvement with pollution has been in connection with soils, it may be well to define what we mean by "soil pollution." D. A. Rennie, Associate Professor, Department of Agriculture, University of Saskatchewan, and his associates gave this definition to the Conference referred to above:

> Any substance that is common or foreign to the soil system which, in addition to the soil either directly or indirectly, adversely affects the productivity of the soil (soil productivity includes both yield parameters and the quality of the food products produced) is termed a soil pollutant (2).

One might be forgiven for believing that with these clear cut and succinct definitions everybody would be aware of pollution and would know where and

Reprinted with permission from *Western Miner*, Vol. 40, no. 12, December 1967, pp. 23-32.
  *University of British Columbia, Vancouver, Canada.
 **University of British Columbia, Vancouver, Canada.

when pollution was taking place. There are close to 3500 million people on this earth and their very being guarantees some alterations to our natural environment, but these alterations are not normally considered as pollution until they reach the "limit of tolerance," and herein lies the rub, who defines the limit of tolerance?

Is the limit of tolerance determined primarily by when some form of life is demonstratively affected or are aesthetic values to be seriously considered?

Actually pollution, or to be more correct, some degree of alteration of our natural environment is part and parcel of our modern world. Whether he smokes a cigarette, drives an automobile, operates a pulp mill or a smelter, or merely burns wood or coal in his furnace, man is upsetting his natural environment. All of which boils down, in many instances, to a matter of opinion as to what is "the limit of tolerance."

It is now our intention to illustrate a few of the problems involved in one small aspect of pollution, namely that aspect of pollution in which British Columbian geologists have particular reason to be concerned, namely, abnormal concentrations of lead, zinc, and copper in our natural environment.

Geological Factors in "Natural Environment"

At the present time, many people appear to believe that water or soil, if they occur in a "natural" or "virgin" state, must be "good," and that if man, by one means or another, changes their composition they become polluted and "bad." Actually, this is by no means the whole story.

The copper, zinc, and lead content of water, soil, and rock varies widely in different parts of the earth's crust. It follows that water and plant samples found in these different parts of the world have widely differing concentrations of copper, zinc, and lead.

Small amounts of copper and zinc are essential to humans who normally acquire adequate amounts of these elements from food and water. However, too much copper or zinc can be just as harmful as too little. A great deal of research remains to be done before we can look forward to knowing just what is best for man as far as achieving an optimum ingestion of copper and zinc from water and food.

Copper and Zinc in Water

In the pollution conference referred to above, a paper was presented dealing with amounts of various metals that might be considered "ideal" and "acceptable" (3) for human consumption. For copper and zinc, the suggested amounts in parts per million—ppm (milligrams per litre) are:

TABLE 1.

	Cu	Zn
Ideal water—"Recommended"—nontoxic	0.2	1.0
Drinking water standards (maximum amounts)	1.0	5.0

Shortly after these data became available, intense controversy broke out about the operation of a mineral concentrator by Western Mines Limited on Myra Creek, which runs into Buttle Lake on Vancouver Island. The little community of Campbell River acquires water from Buttle Lake, and this water, in due course, finds its way into the water supply of Campbell River. Although not qualified to discuss all aspects of pollution, we were interested in ascertaining just how much copper and zinc were involved in various samples of water. Perhaps it should be pointed out that water sampling is one of the recognized tools used by minefinders. On the basis of samples collected at this time the facts were interesting to say the least. Table 2 gives in p.p.m., the copper and zinc contents of various water samples.

The data below need little comment. However, check analyses by another laboratory produced similar results. It would seem that, as far as copper and zinc are concerned, we have more reason to examine modern plumbing than we have to fear pollution from a well supervised and constructed mineral concentrator.

At a later date we learned privately that the water below the Western Mines mill was further tested by placing in it fingerlings for a ninety-six hours' sojourn. All the fingerlings survived.

The data below also serve to introduce another and less appreciated feature of

TABLE 2. Copper and zinc content of "natural" and tap water samples.

Sample Site	Date taken	Cu	Zn
Myra Creek above concentrator	8.2.67	*0.020	0.100
Bridge at tailing pond	8.2.67	*0.020	0.060
Below Myra Falls	8.2.67	*0.020	0.020
Buttle Lake water from tailing disposal area	8.2.67	*0.020	0.040
North end of Buttle Lake	8.2.67	*0.020	*0.010
Campbell River water intake	8.2.67	*0.020	0.004
Tap water B†			
Campbell River Hotel	9.2.67	0.030	0.020
Private home, Campbell River	8.2.67	0.040	0.025
Private home, Point Grey, hot water	?.2.67	0.060	0.020
Private home, Point Grey, cold water	?.2.67	0.040	0.020
Private home, Kerrisdale, hot water	?.2.67	0.200	0.200
Private home, Kerrisdale, cold water	?.2.67	0.500	0.250
Private home, West Vancouver, cold water	6.2.67	1.000	1.800
Private home, University Hill, cold water	3.2.67	0.060	0.500
Prominent Vancouver hotel, cold water	6.2.67	0.400	0.900
Prominent Vancouver motel, cold water	5.2.67	0.400	0.050
Prominent Vancouver club, cold water	4.2.67	1.500	0.150
Prominent Vancouver office building	6.2.67	0.800	0.500

†All samples except those from Campbell River area are from the same North Vancouver watershed.
*Less than.

pollution. We have seen how much water has less than ideal amounts of some elements. Obviously, if we do not obtain minerals from water, we must do so from air or food. Except in some exceptional circumstances, air does not contribute significantly to our supplies of copper and zinc. Can food be counted on to supply the copper and zinc needed by man?

The Copper and Zinc Content of Soils and Food

Because of man's need for more and more food in attempting to cope with the population explosion, agriculturists are taking more and more nutrients, including copper and zinc, from soils. On top of this, modern industrial methods for the refining of food are removing large amounts of copper and zinc, so that foods actually eaten by man are more than likely to be deficient in these elements. Henry A. Schroeder, Professor of Physiology at the Dartmouth Medical School, and one of the world's best known trace element researchers, wrote recently, "Elements essential for grains (and man) are concentrated in the germ and bran (and in all seeds), the carbohydrate endosperm merely providing nourishment until developing roots can get it from the soils. The refinement of flour and rice removes 80-90% of the bulk and trace metals; thus a major source of calories provides what may be less than optimal amounts of metallic micronutrients." Further on in the same article, Schroeder reminds us that Gabriel Bertrand pointed out 65 years ago that "all essential trace elements are toxic in excess" and that "were it not for exquisite homeostatic mechanisms which reject excesses and conserve in the face of deficiencies, men would have to limit his 'intake'."

Thus, if we remember that at least one definition of pollution involves any alteration of natural environment so that it is rendered deleterious to man's use means that we may think of soil, and food products derived thereon, as being polluted either if they have too much or too little of one or more trace elements, perhaps a few examples of the copper and zinc content of some soils and some

TABLE 3. Copper and zinc in some soils.
in parts per million
(Sulphuric acid extractable)

Locality	Cu	Zn
Princeton, B.C. (av. of ten samples)	9	54
Merritt, B.C. (av. of ten samples)	10	33
McBride, B.C. (av. of ten samples)	15	88
Somerset (England) (av. of five samples)	2	250
S.W. Ontario (av. of twelve samples)	8	53
S.W. Devonshire (England) (av. of three samples)	143	947
Manitoba (av. of eight samples)	11	56
Saskatchewan (av. of five samples)	8	54
Sussex (England) (av. of nine samples)	6	94
New Zealand (av. of nine samples)	5	78

vegetables will serve to illustrate the magnitude of the pollution problem that man faces at the present time.

With variations of this nature in soils, it is to be expected that foods grown on these soils will also vary greatly in their content of trace elements. Table 4 provides illustrations of variations we have encountered in lettuces.

With variations such as can be seen in the table below, it is simple to appreciate that pollution may well involve either too little or too much of an element. It just happens that both zinc and copper are essential to our well being. Unfortunately, we are not sure in what form these elements are present in various vegetables so we do not know just how much of what is ingested is eventually absorbed. Consequently, although we know a good deal about how much copper and zinc are present in normally healthy humans, we are not able yet to be sure of what is the optimum amount of these elements which we would like to have in our food.

It will be appreciated that the problem of mineral pollution in soil differs radically from the problem of organic pollution, in that practically all organic pollutants are foreign to normal soil. Mineral pollutants are, at least as elements, already present in virtually all soils. The problem with mineral pollutants is whether or not they are present in a soil in amounts capable of altering significantly, in a harmful manner, the balance of soil composition.

Much more research has to be done before we have the knowledge necessary to know what soils may be considered normal and what soils are anomalous to such a degree that they may be considered to be polluted.

Copper and zinc are known to be essential to human existence: only their absolute abundance, and their abundance relative to one another are important. However, three other elements, silver, mercury, and lead, are not known to have any essential role in plant or animal life, and yet are so ubiquitous in plant and animal tissue that we must be concerned with them because in abnormal amounts they are toxic. You will appreciate that any geologist working in British Columbia, must be interested in these elements because it just happens that here we have one of the favoured parts of the world where these three elements are all concentrated. It was in our search for new deposits of these metals, using biogeochemical techniques, that we became interested in what may be termed

TABLE 4. The copper and zinc content of the ash of lettuce
in parts per million

Locality	Cu	Zn
Kootenay, B.C.	27	1100
Holland Marsh, Ontario	73	650
S.W. Devonshire (a) Eng.	60	1300
Derbyshire, England	6	140
Upper Fraser Valley, B.C.	62	280
Lower Fraser Valley, B.C.	15	500
Cornwall, England	110	490
S.W. Devonshire (b) Eng.	110	6300

medical geology, which would, by definition, include the subject of pollution.

To illustrate what we have learnt about pollution in connection with our mineral exploration work, perhaps we may use lead.

Too much lead does not seem to agree with humans. Nevertheless, we appear to be offending common sense when we offer food for human consumption with lead contents sometimes as much as one hundred times what may be taken as normal.

In Table 5 we list the lead content of some soils which we ourselves have investigated. It should be pointed out that the analytic method we have used does not obtain all the lead in any soil, but it does extract the bulk of this metal: The illustrations that follow are believed to be valid. The soils selected for this report are taken at random merely to demonstrate the variations that are known to occur.

The samples taken from localities 1-6 were, in the opinion of agriculturists, representative of normal soils. The samples taken from 7-10 were all of interest because each locality had attracted the attention of responsible medical men who had published one or more articles drawing attention to some unusual disease pattern in the locality.

Samples 11-14 represent samples taken in our routine geochemical work. No relevant epidemiological data are available, but, understandably enough, we would suggest that medical men might investigate these localities with advantage to all concerned.

However, harmful elements in soils do not always find their way into vegetal matter growing on these soils in the same amounts and proportions that they occur in these soils.

Unfortunately, we have not been in a position to take vegetal samples from the precise localities where the soils were collected, but we do have a few casual

TABLE 5. The sulphuric acid extractable lead in various soils
in parts per million

	Locality	No. of Samples	Average lead
1.	University Area, B.C.	27	3.
2.	Nicola Area, B.C.	16	2.
3.	Cariboo Area, B.C.	16	1.
4.	Princeton Area, B.C.	14	2.
5.	Southern Manitoba	8	4.
6.	Southern Ontario	12	4.
7.	Village C - Gloucestershire, England	10	267.
8.	Village B - Derbyshire, England	9	406.
9.	L — Area, New Zealand	2	115.
10.	Sampford Spiney, Devonshire, England	8	51.
11.	1. Area of suspected pollution, B.C.	8	340.
12.	2. Area of suspected pollution, B.C.	12	164.
13.	Coal mining district, Scotland	1	300.
14.	Coal mining district, Wales	1	200.

samples taken in some of the same general areas. With the intention merely of indicating the possible rewards awaiting more detailed investigations in this field of trace elements and epidemiology, we give, in Table 6, some data we have accumulated in our more recent geochemical studies.

It is necessary once again to point out that it would be folly to draw premature conclusions from the scattered data appearing in the above table.

However, it appears to be certain that pollution in one form or another does affect materially the lead content of both lettuce and potatoes. Until medical men know the form in which the lead is present in these vegetables and what proportion of the ingested lead is absorbed by humans eating these vegetables, it will be impossible to assess the seriousness of each type of pollution. It must be apparent that lead can be introduced into our food supplies by various types of pollution including automobile exhaust, burning coal, and by Divine Providence, if we accept lead mineralization as coming under this heading.

It would probably be unwise to continue much longer for fear of boredom. Possibly, before closing, it might be well to add that virtually all the other trace elements we have investigated, and these include molybdenum, nickel, cobalt, arsenic, cadmium, silver, and mercury, show comparable variations in their concentrations in the rocks, soils, and vegetal matter found around the world. It is not yet possible to say what concentrations and combinations of elements provide a "normal" background for optimum development of plant and animal life. Consequently, it will not always be possible to define the degree of mineral pollution in an area until a good deal of "homework" is performed.

Nevertheless, it should be possible to define some instances of gross pollution, and attempt to correlate the element or elements involved with epidemiological abnormalities. Conversely, it should be fruitful to investigate those areas where unusual prevalences of some diseases are recognized, and try to discover if there is any significant degree of mineral pollution in those areas.

It should be realized that studies of the kind suggested above call for a large degree of interdisciplinary cooperation. This interdisciplinary cooperation is not easily achieved because, in the complex world of today, one of the greatest problems that we face is that of interdisciplinary communication. Any university can testify to the tendency for Arts and Science, Science and Applied Science, Engineering and Medicine, to want to go their own way: lip service is paid to the ideal of interdisciplinary cooperation, but the realization of that ideal is far from easy.

Fashions come and go by fits and starts. Pollution is becoming an accepted word, fashionable alike to idealists and politicians. Perhaps, under the canopy of "pollution control" we may be able to tap enough resources to allow medical men, agriculturists, geologists, and administrators, to explore more fully one aspect of our environment which, if better understood, would allow us to plan for a more healthy world.

Acknowledgments

It must be appreciated that the Geochemical laboratories of the Geology

TABLE 6. The lead content of the ash of edible portions of lettuce and potatoes.
in parts per million

	Locality	No. of samples	Vegetable	Pb	Remarks
1.	Fraser Valley, B.C.	2	Potatoes	28.	In farming area
2.	Okanagan, B.C.	1	"	71.	Close to much travelled highway
3.	Kootenay, B.C.	1	"	47.	Within 300 yds. of lead mineralization
4.	Coal mining district, Scotland	1	"	510.	Heavily industrialized area
5.	Coal mining district, England	1	"	160.	Heavily industrialized area
6.	District (a) Idaho, U.S.A.	4	"	11.	Locality not given to us
7.	District (b) Idaho, U.S.A.	2	"	152.	Locality not given to us
8.	Derbyshire, England	8	"	82.	Area of high multiple sclerosis morbidity
9.	S.W. Devonshire, England	2	Lettuce	505.	Area of high multiple sclerosis morbidity
10.	Isle of Man, England	1	"	14.	Area of normal morbidity
11.	Kent, England	1	"	3.	Area of normal morbidity
12.	Fraser Valley, B.C.	2	"	9.	In general farming area
13.	Kootenay, B.C.	2	"	183.	Within 300 yds. of lead mineralization

Department of the University of British Columbia could not have even attempted their trace element studies, of which the above paper represents only a very small part, had it not been for heart-warming cooperation from a great number of their academic colleagues, from numerous government agriculturists, and various officials from several mining companies.

Over the years, funds for this work have been provided by several mining companies, The Geological Survey of Canada, The National Research Council (Grant No. A 1805), the Multiple Sclerosis Society of Canada, the Defence Research Board of Canada (Grant 7510-16), the Dominion Department of National Health and Welfare with the cooperation of the British Columbia Department of Health Services and Hospital Insurance. Kennco Explorations (Western) Ltd. must be singled out for special mention because, for the past sixteen years they have supported trace element research and permitted digressions into epidemiology to a degree both unusual as it has been welcome.

Since early 1967, the Donner Foundation has provided funds in support of specific epidemiological studies. With new equipment and some staff concentrating on this work, and with happily established cooperation from the Royal College of General Practitioners in Great Britain, we hope to expand our trace element epidemiological studies during the next few years.

Many personal friends and colleagues contributed much time and effort in helping us assemble the multitude of samples on which this paper is based. Our heartfelt thanks go to these unnamed but much appreciated collaborators.

References

1. Canadian Council of Resource Ministers. 1966. Pollution and our environment. A National Conference, Montreal, 31 October - 4 November, 1966. Background Paper D. 26, p. 1-2.
2. Canadian Council of Resource Ministers, *op. cit.* Background Paper C. 22-3, p. 1.
3. Canadian Council of Resource Ministers, *op. cit.* Background Paper C. 22-1, p. 5.
4. Proceedings; University of Missouri's 1st Annual Conference on Trace elements in environmental health, 10-11 July 1967, p. 21.

Trace Elements and Health

Wayne A. Pettyjohn*

Near the close of World War II the badly damaged Japanese industrial machine was attempting to keep production in pace with military losses. Throughout several regions of the country, mining operations for heavy metals were proceeding with great haste. Good industrial waste-disposal practices were largely ignored. Along the upper reaches of the Zintsu River basin, milling wastes from a mine producing zinc, lead, and cadmium were dumped untreated into the river. Downstream, the contaminated water was used by farmers for drinking, cooking, and washing, as well as for the irrigation of rice fields. In a relatively short time, the yield from rice paddies irrigated with the muddy water began to decrease, and many of the rice plants showed evidence of stunted growth.

Sometime later it was recognized by Japanese health authorities that many people in the Zintsu basin, as well as throughout several similar mining regions, were suffering from a strange, sometimes fatal disease that caused the bones to become so thin and brittle that they easily snapped. The very painful affliction became known as "itai-itai," which literally means "ouch-ouch." The cause of the disease was unknown.

In 1960 Jun Kobayashi (6), a Japanese researcher, examined bones and tissues of itai-itai patients. He found in them large concentrations of lead, zinc, and cadmium; consequently, he began to carry out experiments to test the effects of these elements on soil and plants. Chemical analyses showed that the rice plants had selectively absorbed and accumulated large amounts of these metals from the soil that had been contaminated by the irrigation water. The plants continued to absorb dangerous heavy metals from the contaminated soil years after the original pollution source had been removed.

Ingestion of the cadmium-contaminated rice by the local farmers apparently had caused osteomalacia, a bone disease. This disease causes bones to become thin and brittle because of a loss of calcium from the bone; the calcium is replaced by cadmium.

Reprinted with permission from The *Science Teacher,* May 1972.
*Ohio State University, Columbus, Ohio.

Similar studies were made on Tsushima Island, between Japan and Korea, where cadmium, zinc, and lead are mined. Itai-itai was discovered there also, although it had not been recognized previously—perhaps because early symptoms are more easily diagnosed as backache, headache, arthritis, rheumatism, or lumbago. Without some clue as to the actual cause of the disease, it is nearly impossible to diagnose.

Of special significance are the unusually low concentrations of the metallic elements in the contaminated water as compared to their concentrations in soil or plants. Sixty river samples from the Zintsu basin were chemically examined, but most of them contained *less* than 1 ppm (parts per million) of cadmium and less than 50 ppm of zinc. Twenty-three mud samples, however, had a cadmium content as high as 620 ppm and zinc content as much as 62,000 ppm. That is, the sediment commonly contained 10 times more cadmium and 1,000 times more zinc than did the contaminated water! These metals accumulate in living plant and animal tissues and are introduced into a food chain, which often leads to consumption by man.

Pollution and Trace Elements

Many industries throughout the world dispose of wastes containing heavy metals by dumping them into water-ways. Just because a river contains industrial effluent, however, doesn't mean that the water is harmful. In fact, most of the elements found in waste waters are not as harmful as cadmium. Many of the trace elements, those that occur naturally in concentrations generally less than 1 ppm in natural waters, are essential to man's well-being. In many cases, however, there is only a small safety range between requirement and toxicity.

For example, cobalt is a component of vitamin B_{12}, which is essential to life. The lack of minute amounts of copper in the diet results in nutritional anemia in infants, but large concentrations may cause liver damage. Iron is an essential component of the protein in red blood cells, and iron deficiency anemia may create several physiological problems. But when iron is present in concentrations greater than about 1 ppm, the water will taste so metallic as to be unpalatable. On the other hand, even minor amounts of substances such as arsenic, lead, cadmium, mercury, and selenium may cause serious adverse physiological effects in man.

A great deal of national and international concern was generated following the mercury poisoning of several members of the Huckleby family in New Mexico—and from the discovery of mercury-bearing fish in Lakes St. Clair and Erie. In each of the above cases the source of the toxic material was traced, found, and identified. In New Mexico, the source was an organomercury compound used as a preharvest fungicidal seed-dressing for corn. The treated corn was fed to pigs, and the pork was consumed by the family. The toxic effect of the mercury, which remained in the pork tissues even after cooking, resulted in severe brain damage to two children. In Lakes St. Clair and Erie, mercury contamination was traced to factories disposing of their effluent directly into

Lake St. Clair. The subsequent ban on fishing in this lake and in the western part of Lake Erie profoundly affected the commercial fishermen in that area, as well as many sports-fishermen and recreational and tourist enterprises.

Widespread publication of these shocking facts by the news media served to forewarn the general public of the possible toxic effects of some industrial effluents. In a matter of weeks, various state and federal agencies began to look at the concentrations of minor elements in water supplies.

Many people assume that the presence of toxic or potentially toxic trace elements in water and food has only recently been discovered. This is actually far from the case. . . . Reports describing "Minamata" disease, a form of mercury poisoning, began to appear in Japanese journals in 1959, following the death of more than forty people. The problems at Minamata Bay in Japan began as early as 1953 (11). In fact, scientific and medical reports from throughout the world for years have pointed out that many trace elements are common in waste waters; that they tend to be removed from the water and mud by plants and by absorption on silt and clay; and that to determine the presence of such elements it was not enough to test only the water. Those who believe the mercury problems in Lakes St. Clair and Erie are solved are sadly mistaken. The mercury is in the mud on the bottom of the lakes and rivers where it will be slowly released to plant and animal life for decades.

All water contains some trace elements, but their significance in metabolism is a long way from being completely understood in spite of years of work and research by health scientists. Charles Dorfor and Edith Becker compiled a mass of chemical data on raw and finished water supplies from one hundred of the largest cities in the United States: Every sample contained some trace elements (2). In some cases sufficient information is available to set up specific water quality standards, but in others the standards are no more than mere educated guesses that can serve as only temporary guidelines (Table 1).

Arsenic

Arsenic is a well-known poison that occurs naturally in many rocks, minerals, and soils. It may be present in relatively large magnitudes in coal and, when released to the atmosphere as flue dust, may act as nuclei for the accumulation of moisture, which in turn may lead to arsenic-rich rain.

Several industrial processes involved with ceramics, tanneries, chemical, and metal preparation require the use of arsenic, but the manufacture of pesticides consumes the largest amount.

Members of the Kansas Geological Survey have found 10 to 70 ppm of arsenic in presoaks and household detergents (1). Even in the Kansas River the amount of arsenic closely approaches the recommended limit as set by the Public Health Service. Its presence is due to waste-water disposal from municipal treatment plants and septic tanks. Arsenic compounds are common in surface water, owing to the disposal of industrial and municipal effluent and as a result, runoff of rain water and snow melt, which erodes soil containing arsenic

TABLE 1. Public Health Service recommended limit of selected trace elements in drinking water supplies and the concentration of a few trace elements in municipal water supplies at five major cities in the U.S.

Trace element	PHS limit parts per million	Cleveland	Gary	San Jose	Mobile	Wichita
Arsenic	.05	—	—	—	—	—
Barium	1.0	.032	.032	.084	.025	.034
Cadmium	.01	—	—	—	—	—
Chromium	.05	.0035	.002	.001	.0019	ND
Fluoride	variable	—	—	—	—	—
Lead	.05	.0079	.02	.011	.001	ND
Selenium	.01	—	—	—	—	—
Silver	.05	.00023	.00054	.00032	0.00009	ND
Aluminum	None	.25	.096	.049	—	.094
Cobalt	None	ND	ND	ND	ND	ND
Copper	1.0	.006	.0017	.003	.00062	.0012
Zinc	5.0	ND	ND	ND	ND	ND

— not looked for, ND not detected.
C. N. Dorfor and Edith Becker, "Public Water Supplies of the 100 Largest Cities in the United States." U.S. Geological Survey Water Supply Paper 1812. 1962. p. 364.

pesticides. Investigations by the Illinois State Geological Survey show that the uppermost part of the sediment on the floor of the southern part of Lake Michigan contains arsenic ranging in concentration from 5 to 30 ppm (9). The presence of the arsenic is attributed to man's activities in the watershed surrounding the lake.

As indicated, the presence of arsenic in water supplies is not solely the result of pollution. Arsenic poisoning leading to sickness and death among cattle in New Zealand was attributed to drinking water containing large concentrations of arsenic of natural origin (4). Naturally occurring arsenic compounds in waters from the western part of the United States have also been reported.

Small amounts of arsenic are ingested by humans every day. It occurs naturally in food, especially fruit and vegetables, and as a residue of agricultural spraying. It is also present in public drinking supplies.

The toxicity of arsenic to humans is well known. The element accumulates in the body, causing arsenosis. The effects of the poison, when ingested in small amounts, appear very slowly; it may take several years for the poisoning to become apparent. Chronic arsenosis, in its most extreme form, causes death. Arsenic may be carcinogenic, and it is known to affect the liver and heart.

Lead

Perhaps the most toxic element in the heavy metal series is lead. Lead is a cumulative poison. Many children suffer the effects of lead poisoning after eating small flakes of lead-base paint. Lead poisoning may produce mental retardation, palsy, and perhaps severe personality and behavioral problems.

Moonshine whiskey is a source of lead poisoning among unborn or nursing infants as well as adults. The lead dissolves from soldered connections in automobile radiators which are used for distilling. Children born from mothers addicted to moonshine may show retarded growth before birth, as well as spastic and other nervous disturbances after birth. The editors of *Environment* report that lead has been found in 96 percent of the moonshine whiskey sampled in the southeastern part of the United States; of this about 30 percent contained as much as 1 ppm. (3).

Another source of lead in drinking waters—more prevalent in the past than in the present, thanks to the now widespread home use of copper tubing—is the solution of lead pipes. It has been suggested that one of the reasons for the decline of the Roman empire was lead poisoning. Apparently, many of the patricians had water piped into their homes and baths through lead pipes; they drank their wine from lead containers. The constant ingestion of this lead-bearing water and wine may have caused mental retardation of the ruling class.

The major source of lead in the atmosphere is the combustion of lead-based gasoline. It has long been known that emission of lead-containing automobile exhaust along major highways has contributed to the unusually high concentrations of lead in roadside plants. In addition to automotive exhaust, lead arsenate sprays and fumes from certain manufacturing plants, such as smelter fumes, can also introduce large amounts of lead into the atmosphere, soil, and vegetation. Near major highways in British Columbia and in southern England, cereals and vegetables contain 5 to 10 times the normal amount of lead. In both Finland and the United States, food crops grown near major highways contain 4 to 20 times the normal amount of lead (12). These high concentrations are mainly the result of pollution.

Localized excessive concentrations of trace elements led H. V. Warren, a geologist at the University of British Columbia, and his co-workers to point out that "abnormal concentrations of copper, zinc and lead can, in a geographical sense, be related with areas where the over-all mortality rate and the mortality rate for cancer of the stomach are well above average" (13). They also investigated several regions in England and Scandinavia and found that in every area where multiple sclerosis is especially prevalent, garden and farm products contained abnormal amounts of lead.

Two researchers at The Ohio State University found that in all but one of 27 samples of snow collected over a 20-square mile area in Columbus, Ohio, the concentration of lead exceeded the Public Health Service recommended limit for drinking water. Lead concentrations in the January 1970 snow ranged between

0.05 and 1.09 ppm. It was suggested that automobile exhaust was the probable source of the lead (5).

Much of the 12-million pounds of lead shot expended annually by hunters comes to rest on the bottom of shallow lakes and swamps. Eventually the pellets become food for unsuspecting water fowl, causing mass mortality from lead poisoning. It has been estimated that about one million birds die annually from ingesting lead shot.

On the other hand, abnormal concentrations of lead in the environment cannot always be attributed to pollution. Some local soils are known to contain a great deal, because the bedrock from which the soil was derived also contains large natural amounts of lead.

Several common lead compounds may be major sources of this toxic heavy metal in water supplies. Lead acetate is used in the printing and dyeing industry. The insecticide used mainly for the control of the gypsy moth and boll weevil contains lead arsenate, which could act as a major source of soil pollution. Lead chloride is used in the manufacture of lead-base paints and in solder. The high concentration of lead in some sediment samples below some effluent outfalls near oil refineries is probably the result of lead tetraethyl—the substance that also causes lead to occur in large concentrations in the atmosphere through the burning of gasoline.

Zinc

Various types of zinc salts are used in galvanizing and in the manufacturing of paint pigments, pharmaceuticals, cosmetics, and several types of insecticide. The solubility of many of these salts in water accounts for their presence in industrial waste.

Seawater contains an average of only .01 ppm of zinc (7), but soils derived from a zinc-rich rock may contain several hundreds of parts per million. The zinc in the soil may, in turn, be removed by plants, which may concentrate it to as much as 20,000 ppm.

Except at very high concentrations, zinc alone has no known adverse physiological effects upon man. But acid drinks, such as lemonade or fruit juices, should not be made or mixed with zinc-bearing water because certain organic zinc compounds may be poisonous.

On the other hand, zinc is both essential and beneficial in human nutrition. It is a component of insulin and is involved in the utilization of carbohydrates and protein. Some evidence seems to indicate that it aids in the healing of wounds (10). Zinc is not stored in the body and very little is known about its deficiency in man, but zinc deficiency does affect animal reproduction. Some common vegetables may not contain sufficient zinc for dietary requirements and, coupled with the general decline in the use of galvanized materials in plumbing and cooking utensils, the availability of zinc to humans and other animals has been greatly reduced.

On the other hand, zinc is highly toxic to fish and other aquatic animals.

Apparently, the zinc may act as an internal poison, or it may form an insoluble compound with the mucus that covers the gills of fish (8).

Large concentrations of zinc in drinking water are undesirable for several reasons. Zinc produces an objectionable taste; it may cause water to appear milky or, upon boiling, to appear to have a greasy surface scum.

Conclusion

Many different kinds of trace elements in hundreds of different chemical combinations await us in food, water, and air. Some are toxic, some are not; but little is known of possible side effects that can occur following their intake over prolonged periods. Trace elements, by definition, are present in only minute amounts. Yet even small quantities can cause serious health problems. Illnesses caused by them are difficult to diagnose, and it may require years for physiological changes to appear. Once they contaminate the soil they will remain there to be slowly removed by plants and animals for perhaps centuries. Only rarely are food and water supplies chemically examined for them. At least, let us beware.

Class Projects

Interesting class projects might be to attempt to determine the amounts and kinds of trace elements that occur in the municipal water supply or in waste waters released to the environment by local industries. These projects would not be easy for several reasons. First, the chemical determination of trace substances requires both expensive and highly sophisticated instrumentation that is commonly not available to colleges, much less high schools. Second, even the local water-treatment plant operator may have no idea as to the types or concentrations of trace elements occurring in the municipal supply. Third, industries are not likely to give out information concerning either the quantity of waste water discharged or its chemical composition. Nonetheless, projects of this type would tend to point out some of the many problems facing state, federal, and local pollution control and health agencies as well as interested environmentalists.

References

1. Angina, E. E., et al. 1970. Arsenic in detergents: Possible danger and pollution hazard. Science 168:389-390.
2. Dorfor, C. N., and Edith Becker. 1962. Public water supplies of the 100 largest cities in the United States. U.S. Department of the Interior, Geol. Survey Water Supply Paper 1812, p. 364.
3. Editors. 1970. Worst news of the month. Environment 12(3):S-4.
4. Grimmett, R. E. R., and I. G. McIntoch. 1939. Occurrence of arsenic in soils and water in the Waiotapu Valley, and its relation to stock health. New Zealand J. Scientifi Technology 21:138A.
5. Hamilton, W. L., and J. E. Miller. 1971. "High lead concentration in Columbus snow." Ohio J. Science 71:313-316.

6. Kobayashi, Jun. 1969. Special researches on the behavior of cadmium in the Zintsu-River basin. (in Japanese) Japanese Bureau of Science and Technology 39:286-293.

7. Mason, Brian. 1952. Principles of geochemistry. New York: John Wiley and Sons, p. 195.

8. McKee, J. E., and H. W. Wolf. 1963. Water quality criteria. California State Water Quality Control Board Pub. 3-A, p. 140.

9. Ruch, R. R., E. J. Kennedy, and N. F. Shemp. 1970. Distribution of arsenic in unconsolidated sediments from southern Lake Michigan. Illinois State Geological Survey, Environmental Geological Notes 37:16.

10. U.S. Department of Agriculture. 1970. Toward the new. Agricultural Information Bulletin 341, p. 60.

11. U.S. Department of the Interior. Mercury content in the natural environment, a cooperative bibliography. Government Printing Office, Washington, D.C.

12. Warren, H. V. 1961. Some aspects of the relationship between health and geology. Canadian J. Public Health 52:157-164.

13. Warren, H. V., R. E. Delavault, and C. H. Cross. 1967. Possible correlations between geology and some disease patterns. Annual, New York Academy of Science 136:700.

Part Six

Water Pollution and Legal Controls

Regulations governing water use and misuse in individual states are found in their respective constitutions, common laws, and statutory provisions. Water law, at least in part, should be considered a branch of property law. Otherwise technical solutions to water management problems may be doomed to failure because of conflict with the law.

Two major theories of water law are found in the United States; the riparian and the appropriation doctrines. Generally, the riparian doctrine has been applied to the more humid parts of the country while the appropriation concept is common in drier regions. These concepts as described in the first report in this section are often modified to best suit the hydrologic conditions that exist within each state.

The object of the first report in the section is to provide some general background information on the underlying principles of state water laws. The report is not conclusive or for that matter even up-to-date in some instances. Nonetheless, it spells out the basic concepts.

The major doctrines of water law, however, do not reflect the entirety of water pollution legal problems. Other important legal concepts deal with nuisance, negligence, and trespass. In more recent years other legal theories are becoming more commonplace and include, among others, ultrahazardous

Photo opposite courtesy of Henry H. Valiukas, MECCA, St. Paul, Minn.

liability, product liability, antitrust actions, maritime suits, shareholder suits, and the concept of public trust as well. There is much more involved in environmental legal actions than the mere collection and interpretation of technical data.

The National Environmental Policy Act of 1969 is commonly referred to as the environmentalist's Bill of Rights in that "Congress recognized that each person should enjoy a healthful environment and that each person has a responsibility to contribute to the preservation and enhancement of the environment." Under the terms of this act, federal agencies are required to consider the environmental impacts caused by their activities and to submit impact statements to the Council on Environmental Quality.

The Water Quality Improvement Act of 1970 (Public Law 91-224) amended the Water Pollution Control Act of 1948 and replaced the Oil Pollution Act of 1924. It deals mainly with pollution from mining activities, oil spills, and sewage from ships.

Other federal acts that deal with some phase of water or water pollution include the Solid Waste Disposal Act of 1965, Clean Water Restoration Act of 1966, Water Resources Research Act of 1964, Water Resources Planning Act of 1965, Oil Pollution Act of 1961, and the Rivers and Harbors Act of 1899, among many others.

The Rivers and Harbors Act has been only recently rediscovered, expanded in interpretation, and used by the general public as well as the federal government for pollution control. The Refuse Act, which is part of the Rivers and Harbors Act, expressly forbids the dumping of refuse into navigable waters or tributaries of navigable waters of the United States without a permit issued by the U.S. Army Corps of Engineers. A brief outline of several significant features of this act is included in this section. These deal with possible enforcement by citizen action.

Although the Rivers and Harbors Act has been on the books more than 70 years, it has not been effectively used for pollution control. In recent years Congressional hearings have resulted in several recommendations charging the Corps of Engineers to reevaluate their procedures for the issuance of discharge permits. Current congressional ideas are succinctly described in the final article in this section entitled, "Our Water and Wetlands: How the Corps of Engineers Can Help Prevent Their Destruction and Pollution."

Legal Approaches to Water Rights

The legal concepts concerning water, which set the permissive limits of development, have necessarily been founded upon the hydrologic information available at the time they were formulated and thus reflect the incompleteness of scientific knowledge to a degree. The hydrologist can build from scratch in the area of his most profound ignorance. The specialist in water law has a more difficult job, because legal concepts have been developed even when the hydrologic facts and the underlying physical prinicples were not known.

Harold E. Thomas,
in *The Conservation of Ground Water*

Public interest in problems of water law does not arise until critical water situations occur. This is attested to by the fact that the major body of water-use law has developed in the 17 western states, while the larger body of drainage law has developed in the eastern part of the country. These two bodies of law represent the major problems that occur in the field of water—use and damage. This relationship between law and practical needs is an important factor to remember in any analysis of problems of water law.

Numerous conflicts have arisen over rights to use water in the portion of this country normally considered wet—the eastern portion of the United States. But these conflicts usually are smoothed over and forgotten when the water supply becomes adequate again. This never has been the case in most of the western states. From necessity, they have had to adopt statutory law that eliminates the conflict that would have been perpetual.

Public interest in water takes forms other than the formal interventions of the courts and legislatures. As a matter of fact, these interests—judiciary and legislative—usually are the culmination of a great deal of grass roots interest and activity. This interest and activity seem to arise mostly during periods of stress.

Reprinted from *Water Rights in Ohio: Research Report 1*, Ohio Legislative Service Commission, January 1955.

During times of drought or times of flood and immediately after, newspaper and magazine articles on the subject appear; committees are formed and legislators are petitioned for action to soften the damage of future disasters. On some occasions these pleas are answered. They are sometimes answered with good law, sometimes with bad.

The waters with which man is concerned take two major forms: water *on* the ground and water *in* the ground. Each of these may be further divided, so that water *on the ground* takes two patterns: diffused surface waters* and streams. (The latter are referred to in legal parlance as watercourses.) The term "diffused surface waters" refers to waters moving over the surface of the earth but not a part of any defined stream channels. Examples of this are precipitation run-off, flood waters extending well beyond the banks of streams, and water standing in swamps and marshes. Streams, or watercourses, are familiar to all. However, the courts do hold that in order for a body of water to qualify as a watercourse the following requirements must be met: it must be a body of water that flows, usually in a definite channel having a bed and sides, or banks. This flow may be constant or intermittent (1).

Waters in the ground are divided into two categories: subterranean streams and percolating waters. Subterranean streams, in order to qualify as such, must meet the same requirements as surface streams, as described above (2). Percolating waters are those which pass through the ground beneath the surface of the earth without traveling through definite channels (3).

Around each class of waters has grown a body of law, either court-made or statutory. These laws represent the formal action that has been taken to alleviate problems arising from water crises. The law of diffused surface water is germane to the present study only in a very narrow sense. But this damage aspect of water should be considered because certain use problems might, in the future, be solved by utilization of water which was formerly considered to be merely a nuisance. The present law of diffused surface waters has very little to say about the use of such waters.

Diffused Surface Waters

Early in the legal history of this country, diffused surface waters were considered to be a nuisance to be eliminated by any means possible. A landowner could erect dikes, embankments, and similar structures to protect his land from the ravages of flood waters. He could do this regardless of the effect it might have on the property of fellow landowners. If a property owner flooded and damaged the property of an adjoining owner through efforts to protect his own property, the adjoining owner had no recourse. It became incumbent on the owner below to adopt means to protect his property, not only from the flood,

---

*In legal works, e.g., Ohio Jurisprudence and court opinions, diffused surface waters are referred to as surface waters. The terminology here employed is more consonant with present-day scientific terminology and makes for a clearer distinction between diffused surface waters and streams.

but from the acceleration of the flood by his neighbor. This principle of law is called the common enemy doctrine (4). This doctrine has at one time or another been part of the law of almost every state which encountered water damage.

However, in many states, the principles of law on this point have been modified. The next step was the adoption of the so-called civil law rule which holds that an upper landowner has a natural easement through lands of a lower owner. He may make use of the natural drainage channels, gulleys, swales, etc., to protect his property from excess water. This rule of law was somewhat more exact than the older common enemy rule; it spelled out more specifically the rights and privileges of landowners than did the other.

However, the civil law rule has in it some features which worked injustices on some landowners. The natural drain facilities were employed to conduct water away. An owner could dig trenches or other devices to conduct water into the natural drains and so increase the natural flow to the point of serious detriment to the lower owner. The civil law rule, strictly applied, provided no recourse for the injured party. In attempting to remedy this situation some of the states invoked the legal maxim, "so use your property as not to injure another," the reasonable use rule. This rule allows for recovery in some cases (5).

These few principles cover most of the law relating to diffused surface waters. To be sure, many of the states have enacted statutes facilitating drainage of lands. But, they are based primarily on the premise that this water is a nuisance and should be disposed of as quickly as possible. That this water could be used for beneficial purposes would seem to be a fairly obvious conclusion, but there is, practically speaking, no law concerning the use of diffused surface waters.*

Ohio once followed the common enemy doctrine respecting diffused surface waters, but the civil law rule eventually was adopted. This class of waters is legally defined in Ohio as those waters diffused over the surface of the ground, derived from falling rains and melting snows. They are diffused surface waters until they reach some well-defined channel, then becoming part of the water-course.

It is a well-settled rule of law in this state that upper owners have natural easements through the estates of lower owners so that diffused waters may drain naturally (6). If property is damaged as a result of natural drainage, the injured party has no recourse. However, if this damage results from "unreasonable" diversions into natural drainage channels, the offender is liable for payment of damages (7).

With respect to flood waters—a special instance of diffused surface waters—it is held that property owners may protect their property against the ravages of floods. In exercising this right, however, they may not cause injury to neighboring property without becoming liable to damage suits (8).

The law of the State of Ohio with respect to diffused surface waters has been

---

*In conversation with the writer and others, Mr. C. E. Busby, a leading authority on the law of water and water rights, stated that he knew of only five cases in the English-speaking world which deal with the use of diffused surface waters.

expanded and extended with the adoption of the drainage laws.* As in other states, these laws, both common law and the statutes, assume that diffused surface water is a nuisance to be disposed of as quickly as possible. Examination of cases and the statutes discloses no cases or rules of law having to do with the use of such waters. Section 6131.59 of the Revised Code might be said to provide for use of water in drainage ditches. That section provides:

> When an improvement consisting of a ditch, drain, or watercourse has become the outlet of agricultural drainage, and it has been established and constructed for seven years or. more, it shall be deemed to be a public watercourse notwithstanding any error, defect, or irregularity in the location, establishment, or construction thereof, and the public shall have and possess in and to any such watercourse which has thus been constructed, or used, for seven years, the rights and privileges which relate to and pertain to natural watercourses, but the same shall be subject to any improvement upon petition as provided in sections 6131.01 to 6131.64, inclusive of the Revised Code.

Presumably, then, the person who owns lands riparian to such a watercourse might use its waters as a matter of riparian right. It might be said that the law of diffused surface waters is concerned almost exclusively with the damage aspects of water.

The law that developed around diffused surface waters has two major objectives. The first was to provide means of protecting landowners against undue damage from the activities of other landowners. The other was to provide legal means for protection of land against the ravages of water. The latter is more true of drainage statutes than of the common law principles.

In accomplishing these aims, the laws might well have been successful. However, the approach of the present law of diffused surface waters is totally negative. It overlooks the possibility that the water might have positive value. The law does not encourage consideration of this possibility. To condemn this law for what it fails to do would be unjust. But the point to be considered is that use might be made of this water.

Streams

An elaborate body of law surrounds the category of water known as streams or watercourses. This is true because streams are more easily understood and more readily used than diffused surface waters and percolating ground waters. It is easy to see the effect of use of such waters, or it is very easy to ascertain that a stream has been polluted beyond reason. Thus the law of streams is more detailed and more exact than the law regarding other classes of waters.

The law of streams has two major aspects: the riparian rule and rights acquired by appropriation. The law of streams applies to both underground and surface streams. It is a presumption of law that all underground water is

---

*Examined in detail in another commission report.

percolating until the existence of a stream is proven. After such proof, the same law which applies to surface streams in any given state applies to subsurface streams (9).

The Riparian Doctrine

This common law rule of decision was part of the early law of the original states. It was not a part of the English common law. The original doctrine of riparian rights was taken from the Code Napoleon early in the nineteenth century, which in turn had been developed from the Roman civil law. Full discussion of the historical aspects of early water law is not essential to this study but can be found in an article by S. C. Weil in Volume 33 of the *Harvard Law Review* of 1919. The language of the article of the Code Napoleon on which our riparian doctrine is based should be noted, however:

> Article 644. He whose property borders on a running watercourse, other than that which is declared an appurtenance of the public domain by Article 538 under the title of Classification of Things, may supply himself from it in its passage for the irrigation of his properties. He whose estate such water crosses is at liberty to use it within the space which it crosses, *but on condition of restoring it, at its departure from his land, to its ordinary course* (10).

This article of the code set the basis for a water policy that, on the face of it, seemed ideal. In later constructions, equality of rights was insisted upon and stressed. This was in keeping with the then popular spirit of liberty, equality, fraternity. Practically, however, this policy presented difficulty. In practice, it becomes a non-use, monopolistic one (11). If every owner had equal rights to the waters of a stream no one could use it. For example, following this theory strictly, if the lowermost owner on any given stream wanted to have the undiminished flow of a stream, none of the upper owners could have used any of the water. And the lower owner didn't have to use the water in order to maintain his right. He could simply demand that the undiminished flow come past his property.

Recognizing the difficulty inherent in the civil code rule, Justice Joseph Story of the United States Supreme Court and Chancellor James Kent of the New York court, in ruling on water cases circa 1825 (12), developed the French rule beyond its original construction. This construction of the riparian doctrine has been best summarized by Busby:

> . . . one who owned lands touching upon a stream was entitled to have the full flow of the stream come by his place undiminished in quantity and unimpaired in quality (Code Napoleon), except that each landowner was entitled to make reasonable use of the water upon his own lands, provided: (1) that he returned the stream to its natural channel before it left his lands, so as not to deprive succeeding lower riparian owners of their rights, and (2) that use on his own land was reasonable in respect to the corresponding rights of other riparian owners lying below him (rule of reasonable use) (13).

This principle of law was the rule of decision used by most of the states prior to 1850; it still is the law, exclusively, in all of the states east of the Mississippi River and in the tier of states just to the west. It permitted use of water in quantities sufficient to satisfy most domestic needs. However, it did not progress far beyond the old principles of the Napoleonic Code. The emphasis on non-use still was present. In the states where water has been abundant, the principles of this law have not been developed much beyond the original pronouncements of Story and Kent.

Under the riparian doctrine, the right to use water is considered real property, but the water itself is not property of the landowner (14). The right is a natural right which is an appurtenance of the property. It is a property that enters materially into the value of an estate; it may be transferred, sold, or granted to another person.

While it is generally held that such rights are natural rights, they are subject to state regulation. Rights of individual owners are subordinate to the public interest. This is true in the case of property rights which have semi-public character and elements of danger. Thus, the states exercising their police power, may subject water rights to reasonable regulation (15).

According to Ohio law, watercourses are defined as water flowing in definite channels having a bed and sides, or banks, and discharging themselves into other streams or bodies of water. The flow may be continuous or intermittent (16). Furthermore, flood waters are normally considered to be part of the watercourse. A flooded watercourse is considered to be at high channel, and the flow at regular levels is considered to be at low channel (17).

Also to be considered with the law of streams are the laws related to the use of waters in lakes and ponds. Though technically these bodies of water are different from streams, riparian rights and littoral rights* are treated as the same in Ohio law.

Following the pattern of most of the common law states, Ohio waters are governed by the law of riparian rights. This right to have the water of the stream come to the riparian owner in its natural flow and in its natural state of purity (18) is a right inseparably annexed to the soil (19).

The riparian owner is entitled to make use of the water flowing by his land to satisfy any reasonable needs for the full enjoyment of his property (20). However, he should not make withdrawals which would interfere with the rights of other riparian owners. Should he do so, he can be liable for payment of damages. In this instance, however, the complainant must be able to prove material damages (21). It is a debatable point of law as to whether destruction of scenic qualities of properties by withdrawals for reasonable uses constitutes a damage (22).

Reasonable use, according to Ohio law, refers to those uses of water which are necessary to the sustenance of life and enjoyment of property. A hierarchy of

---

*The term littoral rights refers to those rights arising from ownership of lands contiguous to the seas or large lakes. The term littoral is from *littus*—belonging to the shore, *Bouvier's Law Dictionary*.

uses has been developed by the courts. First priority is given to water used for drinking, cooking, and stockwatering (23), second to water used for industrial and irrigation purposes; and last, to water used on non-riparian** lands (24).

(In situations where there is conflict among owners over amounts of withdrawal for reasonable uses, it has been determined that when the conflict is over domestic uses, in times of short supply, all owners must bear the burden of the shortage equally)(25). When there has been a conflict between an upper owner for domestic uses and a lower owner for a secondary purpose, e.g., water power, the lower owner has had to forgo his right. And when the conflict is between lower and upper owner for the same secondary use, then the upper owner has had a claim senior to that of the lower owner (26).

Similar rights accrue to persons owning lands contiguous to lakes and ponds. Of course, such a body of water which lies completely within the acreage owned by one person is the property of that person to do with as he pleases (27).

In addition to the rights of riparian owners, certain responsibilities are incumbent upon them. These responsibilities are logical results of the law of riparian rights. If every owner is entitled the full flow of the stream diminished only by the necessary withdrawals of upper owners for reasonable purposes and to the water in its natural state of purity, it logically follows that a riparian owner may not obstruct or pollute a stream (28).

It seems that a riparian owner might have the right to obstruct a stream for a short period of time for some very just and reasonable cause (29). Questions of definition of just and reasonable are questions of fact to be decided in each instance (30).

On a non-navigable stream, a riparian owner could build a dam and direct the water for his use so long as he returns it to its natural channel before it leaves the limits of his property, and provided that it is not diminished in quantity and quality. Violation of these limitations makes the offender liable to the injured parties.

Just as a riparian proprietor may not obstruct a stream unreasonably, he may not pollute a stream. They are not to be polluted by domestic and farm wastes nor by industrial and mining discharges (31). Pollution presents something of a special case because the carrying away of wastes has long been a legitimate function of water with its capacity for the dilution and transportation of waste materials. This factor has been so closely connected with issues of public health, safety, and welfare that pollution-control statutes have gone far beyond the common law principles.

In Ohio riparian law has not been sufficient to meet all problems of pollution. Streams are polluted on a large scale by cities and industries. To attempt to control this pollution, Ohio has had to rely more on statutory law than on riparian law.

(The riparian doctrine is not a policy for water use. It is a rule of decision for

---

*Non-riparian lands are those lands not contiguous to a stream and removed from the drainage areas of a stream. Conversely, riparian lands are, at minimum, those lands which abut on a watercourse.

settlement of disputes among land owners. It was formulated during a time when abundant water supplies were taken for granted, the population was small, and the great advances in agriculture and technology had not yet occurred. The period was characterized by an economy which prompted Ohio's attorney general to say " . . . the frame of reference for the water law was the cabin on the bank" (32).

Riparian law was formulated at a time when the science of hydrology was extremely imperfect. Early court opinions frequently pointed to the inadequacies of the science and, perhaps rightly, used this as a major justification for non-intervention in water matters as they stood. However, the science has been greatly developed since the early 19th century and much more is known of the behavior of water. It now is understood, for instance, that all waters—diffused surface waters, percolating waters, and streamflow—are a part of the same cycle. Yet, different laws apply to each stage in the cycle.

The riparian law, as a rule of decision, is not even fully effective in the achievement of its own ends. A judicial decision on a water right holds only so long as another judge does not interpret the law or the facts differently. And the landowner never is certain of his rights until a judicial determination has been made. Under this system, a landowner's investment is not always secure, to say nothing of investments he might make in such things as irrigation and pumping equipment.

In the light of modern knowledge and modern economic conditions, there is still a more serious defect in this kind of law. It presents no basis or framework for efficient water management. As court-made and self-enforced law, it can not provide such. The "full flow" requirement of the riparian doctrine tends to encourage waste or non-use, which is undesirable in times when areas are faced with critical water shortages.

A state water management program can not be expected to evolve from a common law system. As an answer to this problem, the western states have adopted statutory systems of water law and water rights which incorporate management features. These systems are considered next.

The Appropriation Doctrine

In the western part of the United States, the needs and the laws have been different. First of all, this area has been consistently dry so that the water supply remains relatively short. This part of the country has a tradition somewhat different from other areas, in that it has been under a succession of Indian, Spanish, and Mexican regimes. These people had different customs with respect to natural resources. The early unwritten law with respect to gold and silver "finds" closely parallels the present water laws of the West. With respect to gold, the procedure, had it been formalized into law, would have read something like this.

> He who first discovers gold, if he develops his claim within a reasonable period of time, and continues his development diligently, may have the benefits of his labor. If he fails to develop and work them diligently his rights are forfeited.

When questions of water rights arose, the same rule prevailed (33). This procedure and its rule acquired the name "appropriation doctrine." As the territories became states, some of them adopted the law of riparian rights, but some of them rejected it even to the point of repudiating it in their constitutions. Some of the western states used the constitution as a means of declaring state water policy, and some even employed the constitution to establish the machinery for state administration of appropriation law. Some western states, employing common law, adopted certain features of appropriation law in addition.

Appropriation statutes follow, generally, the mining-camp principle illustrated above. Perhaps the best definition of appropriation law is this description from the *Report of the President's Water Policy Commission:*

> The appropriation doctrine . . . rests on the proposition that beneficial use of water is the basis, measure and limit of the appropriative right. The first in time is prior in right. Perfected only by use, the right is lost by abandonment. . . . An appropriative water right is not identified by ownership of riparian lands. . . . Its existence and relationship to other rights on the same stream are identified in terms of time of initiation of the right by start of the work to divert water coupled with an intent to make beneficial use of it and the diligence with which the appropriator prosecutes to completion his diversion works and actually applies the water to beneficial use (34).

Although details vary greatly, this statement is the essence of water law in the appropriation states. Throughout appropriation law, the elements of beneficial use and water conservation play important roles, because in these states water has been relatively scarce.

Statutory Water Law in the West

Generally speaking, the states of poorest water supply manage their water most closely. Some of these western states, including Colorado, New Mexico and Arizona, never have recognized riparian law. Other western states adopted the riparian doctrine as a rule of decision but also passed appropriation statutes. As water problems became more acute and conflict between the two systems developed, the riparian doctrine usually yielded to the statutory method.

These state systems merit an examination to show the philosophy of government involved in the appropriation system; to illustrate the administrative procedure, and to show the contrast between the two systems.

Of the 17 states having statutory water law, nine have appropriation law exclusively. The other eight have both riparian law and appropriation law working together with varying degrees of compatibility. In Arizona, Colorado, Idaho, Kansas, Montana, Nevada, New Mexico, Utah, and Wyoming, appropriation law is exclusive. Each of the states differs from the others in mechanics of administration, so that a consideration of each is in order.

In *Arizona*, the constitution provides that "the common law doctrines of riparian rights shall not obtain or be of any force of effect in the state" (35). It states further that "all existing rights to the use of any water in that state for all useful or beneficial purposes are . . . recognized and confirmed." To implement this policy, the State of Arizona has adopted a comprehensive water code. In addition to reiterating the state's policy regarding water, the code declares that all water of the state from all sources of supply is subject to appropriation. Preferred uses of water are (1) domestic uses, (2) municipal uses, and (3) irrigation.

To implement administration of the water policy, jurisdiction was given to the Office of the State Land Commissioner. Applications are handled by this office. The commissioner is given authority to establish water districts and appoint a commission for each, to aid in the administration of the program. Further, the commissioner has authority to make rules and regulations and has enforcement powers. Also he has a quasi-judicial function to interpret and make conclusions of law. His determinations are filed in Superior Court. Water proceedings initiated in the courts are referred first to the commission for determination.

A person desiring to appropriate water in Arizona applies to the commissioner. If the rules of the commission are met, a certificate is issued granting the appropriation; the right is lost only when it is not exercised for five successive years. An exception to this procedure is an appropriation for a hydroelectric power development of 25,000 horsepower or more. In this case, an act of the legislature is required; the right is then leased to the appropriator for a period of not more than 40 years with a renewal option provided (36).

This may seem to be a rigid program. But when it is remembered that Arizona has an average annual precipitation of less than 15 inches, it is apparent that this strict management was born of necessity.

Colorado, another of the exclusive appropriation states, follows much the same pattern as Arizona. The state constitution declares all water of natural streams to be public property and subject to appropriation (37). It states further the principle of the superiority of prior appropriative rights and the preferred uses of water—(1) domestic, (2) agricultural, and (3) manufacturing (38).

Responsibility for administering the law of water is vested in the state engineer. He can make and enforce rules, but adjudications are handled solely by the courts.

Applications to appropriate are made to the engineer's office. He issues a permit for the necessary construction. After it is completed, inspected, and approved, the claim is accepted and filed with the appropriate governmental officers (39).

In *Idaho* the mechanics of appropriation are somewhat different. However, the constitutional provisions are about the same as those of Arizona and Colorado, with the exception that it is declared that priority of appropriation gives the prevailing right except in times of short supply, when domestic uses take preference over all others (40).

The Department of Reclamation has the responsibility for executing the law of water. It may make surveys and develop rules and regulations which are consistent with the hydrologic facts. Adjudications are made exclusively by the courts.

Idaho differs from other appropriation states in that it provides two methods of appropriating: the constitutional method and the statutory method. Under the former method, appropriative rights may be acquired by anyone simply by diverting water and putting it to beneficial use. Under the latter, or statutory procedure, application is made to the Department of Reclamation. After necessary construction is completed and the department's requirements are met, the appropriator is given a license affirming his right. In either case, the right must be exercised or it is lost after five years of non-use (41).

This constitutional method of gaining appropriative right is an unusual feature, and it is difficult to assess its merit as compared with the statutory method. There seems to be one advantage in following the statutory procedure. Under this procedure, the priority of right begins on the date of application to appropriate, while under the constitutional method, the right begins on the date the water is put to beneficial use (42).

Idaho presents an interesting situation from a legal point of view, but, despite its mechanical differences from the rest of the appropriation states, the emphasis still is on beneficial use.

Kansas is of interest to most students of the law of water—not because of the nature of its program, but because of the recent adoption of its appropriation law. Kansas adopted riparian law as the rule of decision and a little later adopted appropriation law for water to be used for irrigation. The scope of appropriation law was expanded little by little, and a great deal of litigation resulted, testing appropriation law as against common law. As late as 1944, the Kansas Supreme Court, in a *quo warranto* action (43), issued an opinion holding that the common law was still the law of the state except when modified by statute. Thus, common law claimants, in many cases, could recover damages against appropriation claimants. The next year a complete appropriation code, also governing ground waters, was adopted. The supreme court has upheld this code as constitutional (44).

The law vests authority for administration of the act in the chief engineer of the State Board of Agriculture. It is his duty to decide, after study of the facts, which existing rights are to be recognized and thereafter to administer the new appropriations. Determinations of rights are made by him, subject to appeal to the district court.

The act holds that all waters of the state are subject to appropriation. The following is the order of preferred uses: domestic, municipal, irrigation, industrial, recreational, and water power uses. Water rights acquired through appropriation are lost by non-use (45).

The law in Kansas seems to have worked fairly well. This is attested to by the fact that, in its ten years of existence, it never has been amended (46). There are some questions of law and fact that yet need judicial interpretation, but they seem not to be of major consequence (47).

Montana has a system of appropriation quite unlike that of the other appropriation states. It is different in that it has no centralized state administration of water use; all of the powers of enforcement and administration are in the county courts. The State Conservation Board acts much as a fact-finding agency, though it may initiate proceedings in the local courts to correct any situations which it decides are not in the public interest (48).

Appropriation is accomplished in this fashion: to appropriate waters from an unadjudicated source (any in which no determination of rights has been made), the appropriator merely files notice with the county court and diverts the water. However, to appropriate from an adjudicated source he must hire an engineer to survey the proposed diversion site and its relation to other users' diversions, and then he must file a petition in the county court so that the terms of his right may be decided. After weighing the facts of the case, the court issues a decree granting or refusing the right.

Appropriative rights are lost only by abandonment. There are no statutory provisions for the future. Thus, in each instance, a court must judge the facts in the case (49).

Nevada does not present any great departure from the principles followed in most other appropriation states. Appropriations are made through the state engineer. He administers the law. He makes determinations which must be entered in the courts; these have the effect of complaints in a civil action. All waters are subject to appropriation with the exception of wells for domestic use which have a draught of less than two gallons per minute. Rights are lost by five consecutive years of non-use (50).

In *New Mexico* the situation is exactly the same as Nevada with two exceptions: (1) there is a statutory four-year non-use forfeiture and (2) adjudications are made in the courts, any state action being initiated through the Office of the Attorney General (51).

The State of *Utah* follows a procedure substantially similar to that of New Mexico (52).

For the State of *Wyoming,* the same statement may be made with the exception of a change in name. Instead of the state engineer, the State Board of Control is the administrating agency, with the engineer acting as a member of its staff (53).

Wyoming completes the list of the states operating exclusively under statutory appropriation law. In the eight other western states—California, Nebraska, North Dakota, Oklahoma, South Dakota, Texas, and Washington—the appropriation doctrine and riparian doctrine co-exist with varying degrees of compatibility. The question of how the doctrines co-exist deserves special examination.

In *California* a system of limitations is imposed on each type of law. For example, only certain water is subject to appropriation: water on public lands and the surplus from private lands. The riparian owner must adhere to a strict interpretation of reasonable use, and his use must be both reasonable and beneficial.

There was much litigation over the relative rights of riparian and appropriative users. This led to a constitutional amendment in 1928 which restricted riparian rights to reasonable use and reasonable methods of diversion. Beyond that, riparian rights are not recognized. The riparian owner's rights to this extent are protected by law. These rights can be lost through prescription and judgment, but the owner must be compensated. All surplus above the riparian owner's rights is declared part of state property and is subject to appropriation(54).

In *Nebraska* the riparian doctrine is part of the law of the state. However, the Supreme Court of Nebraska has consistently reduced its importance as the basis for deciding disputes. Two decisions, given on the same day, held that the common law applied except as limited by legislation (55). However, two years later, in a similar dual decision, the court ruled that unless the riparian owner had been making actual use at the time of appropriation by another person and had proof, he had no alternative but to seek damage. His advantage in location was lost to an advantage in time (56). While the riparian owner's advantage in location and his riparian rights are to a limited extent yet recognized, he must go through the formality of appropriation procedure to establish that his diversion is for reasonable and beneficial use (57).

North Dakota has a situation whereby both systems operate. There has been little litigation in the state testing the strength of the two systems. Section 210 of the North Dakota Constitution provides that all flowing streams and natural watercourses shall forever remain the property of the state for mining, irrigating, and manufacturing purposes. The state Supreme Court in *Bigelow v. Draper* (6 N. Dak. 152) held that constitutional provision did not, by declaring water to be state property for designated purposes, divest the state of the system of riparian right. In *Sturr v. Beck* (133 U.S. 541) the United States Supreme Court held that the riparian doctrine was the law of the state. There have been only a few cases regarding this issue in the history of North Dakota. In each it was a sharp conflict which could be settled with the common law principles. The full test is yet to come (58).

In *Oklahoma*, at least until very recent times, the situation was unclear. Both systems operate, but they have operated in rather exclusive spheres. No decisions testing the strength of each in opposition to the other has been rendered. The situation is analogous to that of North Dakota.

". . . the riparian doctrine, in *Oregon*, appears to be little more than a legal fiction," says Wells Hutchins in the *Report of the President's Water Policy Commission* (59). The appropriation doctrine is the predominant of the two systems by far. Riparian owners are limited to beneficial use, and they are not guaranteed undiminished flow (60). At the time the appropriation statutes were written, the vested rights were recognized only if waters were being put to beneficial use at the time of enactment or shortly thereafter. Riparian rights were further recognized to the extent that one claiming riparian lands could claim, for beneficial use, water rights either appropriative or riparian, but not both. In summary, it is fair to say that the mass of judicial decisions has converted Oregon into what is an essentially appropriation state (61).

In *South Dakota* the situation is essentially the same as that of North Dakota. The two states have a common background for law, i.e., the laws of the Territory of the Dakotas before the division into states.

In *Texas*, riparian rights seem to prevail over appropriation rights. The language of the appropriation statutes always has protected vested riparian interest. When the two have come into conflict, the riparian doctrine has been the basis for adjudication. The riparian doctrine was affirmed by the courts as early as 1863 and as late as 1934 (62).

The major aspects of the dilemma are avoided in the State of *Washington* because the appropriation statutes apply only to waters flowing through the public lands. Vested riparian rights are protected by law, and when title is acquired to lands from the public domain, the water no longer is subject to appropriation (63).

It is evident that the co-existence of the two doctrines is a difficult principle in application. Nowhere are the two systems co-equal. In some states, the status is unclear because there have been few test cases.

Comparison of the Two Systems

Comparison of the major features of appropriation law with those of riparian law shows that, under the statutory system, a framework is provided to correct the deficiencies of the common law system.

In the determination of rights, the issue is settled by a state agency, and the appropriative right is a permanent, exclusive right, guaranteeing security of investment. Frequent and protracted court battles rarely occur under appropriation law.

Appropriation statutes are the implementation of a beneficial use policy. Waste, non-use, and non-beneficial uses can be discouraged. While the appropriative right in some instances is monopolistic, it is so only in terms of beneficial use.

Water agencies in the appropriation states are empowered to make and enforce rules. An administrative agency with such authority can frame its regulations in terms of need, and these rules can be based on scientific principles. With scientists participating in making a policy, there is greater likelihood that the law will keep up with the knowledge of hydrology.

It appears that comprehensive and effective management of watercourses probably could occur only when a state adopts the major principles of the appropriation system.

Percolating Waters

The earliest principle of law on percolating ground waters is the so-called English rule which is based on the legal maxim, *"Cujus est solem ejus ad collum et ad inferus."* (He who owns the soil owns from the heavens to the depths) (64). Following this principle, it is held that the property owner may do as he sees fit with his property; his right to use any of its resources is absolute. Strictly construed, this means that a man on his own property may dig a well or many wells, or dig a quarry, or drill for oil, or mine coal without any consideration of the effect any of these activities might have on the water level of his neighbors. If wells or springs go dry as a result the offending party is not liable for damages. The reasoning behind the establishment of such a rule reflects an imperfect knowledge of geology and hydrology. Little was known about the behavior of ground water, and thus it was difficult to develop rules regarding it. Judge Jacob Brinkhoff, in his opinion in *Frazier v. Brown* (65), illustrated the climate of judicial opinion on this matter when he said, concerning ground waters:

> Because the existence, origin, movement and course of such waters, and the causes which govern and direct their movements are so secret, occult, and concealed, that an attempt to administer any set of legal rules in respect to them would be involved in hopeless uncertainty, and would be, therefore, practically impossible.

A substantial body of precedent has developed around this principle of law, and it has been the most widely accepted rule of decision in ground water cases.

Ohio's law relative to rights to the use of percolating water has not gone beyond the English rule. Percolating waters are legally defined as those waters which pass through the ground beneath the surface of the earth without flowing in definite channels (66). All ground waters are presumed to be percolating (67), unless a litigant can convince the court that an underground stream exists.

It seems a well-settled rule of law in Ohio that a landowner may make full use of the waters percolating beneath the surface of his land. He is not liable for damages if neighboring owners lose their water as a result of his operations (68). Owners losing water in springs or wells have no recovery on the grounds of negligence or maliciousness. At one time maliciousness seemed to be a ground for recovery of damages. In *Wyandot v. Sells* (6 ONP 64; 9 O. Dec. Rep.) damages were awarded on the basis of maliciousness, but later the court reversed itself. Since this was the only case found concerning itself with this question of maliciousness, it seems fair to conclude that the English rule still holds strictly (69).

There seems to be one situation in which a landowner might recover damages for loss of percolating water sources—when sub-surface geologic formations which make for fruitful yields of wells and springs have been destroyed. An opinion was delivered in 1923 (70) covering two similar cases where, including other damages, wells had been destroyed on lands leased to coal companies for mining operations. The courts held that this was an instance of damage and found for the plaintiffs. Apparently this is the only instance in which recovery might be had for the loss of percolating waters.

Although the landowner is free to use percolating waters in any manner he chooses without regard to the destruction of supplies of others by withdrawal, he is charged with the responsibility not to pollute percolating waters. The landowner is liable if he allows accumulation of deleterious materials on his land and these contaminate waters of others. This holds true for materials piled on the ground or for waste materials stored in improperly constructed cesspools (71).

In summary, Ohio considers percolating waters to be the property of the overlying owner, and they may be used as the owner sees fit. His sole duty is to refrain from polluting percolating waters.

In some states the rule has been modified. There were instances of hardship that could be traced directly to an offending construction. For example, a dried-up well could be traced directly to a deeper neighboring well drying up a nearby spring.

There is another rule of law designed to correct some of the inequities involved. It was based on the legal maxim *"sic utere tuo et alienum non laedas"* (so use your own as not to injure another). This came to be known as the reasonable use rule. Under this rule, the landowner should use the water beneath his soil reasonably, having due regard for the rights of others. Implied in this reasonable use rule is the theory that an owner, while he has the right to use the water in his soil, should not use it wastefully nor negligently, nor in such a way as to destroy maliciously the property of another. Under this rule, damage can be recovered in cases of loss of water due to any of the named factors.

This rule has had a somewhat more limited application than has the English rule. It does help to resolve some of the inequities inherent in the older rule, but it still fails to resolve all conflicts and affords no effective basis for management of ground waters.

California went somewhat further than most states in establishing the principle of correlative rights. In a case involving a dispute between owners of land overlying a large artesian basin (72), the courts decided that the English rule did not fit that state's needs and discarded the rule. The law holds that owners have the right only to so much water as will satisfy the needs for beneficial use on the property, the surplus being available for appropriation. The California rule provides further that, in times of shortage, each owner must bear his share of the shortage, and none can take advantage of a favorable geographic position.

Several* of the appropriation states abandoned the English rule entirely and required that all ground water be subject to appropriation. They have rigid requirements for well construction and amounts of water to be used. The exception to this is that constructions for domestic use, usually not having a large draught of water, are exempt from the regulations. Most of the states requiring appropriation of ground water have adopted this principle in recent times, after ground water crises.

Progress in developing adequate laws for managing the use of ground water has been slow. This situation can be attributed to the fact that the paths of ground water are not visible and are somewhat mysterious to the person who is not a geologist. Another factor is the present conception of private property—people who own land like to feel that it's theirs to do with as they please. This attitude is deeply ingrained, and many proposed laws are directly opposite.

The laws of ground water, especially the common law principles, have inherent difficulties in them, not the least of which is that they have not kept

---

*Those states are Arizona, Idaho, Kansas, New Mexico, North Dakota, Oklahoma, Oregon, Utah, and Wyoming.

abreast of scientific knowledge. To scientists, the paths of ground waters are not so "secret and occult." It now is possible to administer efficiently a set of rules for use of percolating waters.

The following diagram (figure 1) summarizes what has been said about the classes of water and about water laws. There are two major aspects of water law: use and damage. Use refers to those rules of law which apply to water users and property owners. Damage refers to those rules of law which apply to injuries inflicted by water or those injuries resultant from pollution. Both of these have antecedents in the common law, though in some jurisdictions they have been important enough for embodiment in statutory law.

Criticism of the Common-Law Approach

Water rights law is incidental to the physical and economic realities of water itself. Water rights law does and should conform to those realities. And where the law itself is inconsistent with real needs or practices, review and possibly revision of the law are needed. Following are several aspects of the common-law approach which are, in some measure, inconsistent with the physical and economic facts.

Classification of Waters

The non-statutory legal structure divides water into more or less arbitrary classes, without recognizing that all water is part of the same cycle. Thus, one body of law applies to ground waters and another to streams, without recognizing that the two flow into one another and therefore the use of one often affects the other. There are instances in which a major user of ground waters actually is using water which seeped out of a stream bed. When the stream flow is impaired, his ground source may be seriously impaired as a result. Yet he has no recourse if an upstream user impairs the stream flow, while a downstream riparian owner would have recourse.

Beneficial Use

Effective water management requires that the most essential uses have adequate water sources at all times and that less important uses generally should be met only after these are provided for. Under non-statutory law, the only distinction made between a more essential use of water and a less essential use is one which has been devised by the courts as a rule of decision in civil suits. Thus, where the local supply of usable water is less than ample, the assurance of water supply for the absolutely essential uses might depend upon civil litigation, after the fact, between individual parties. This affords no means of preventing excessive use for less essential purposes before the fact, and it relies upon the relatively unpredictable mechanism of lawsuit.

Waste and the Impairment of Water Sources

It is wasteful to permit large volumes of water to run off in the streams without maximum opportunity for use. The non-statutory approach vests exclusive rights

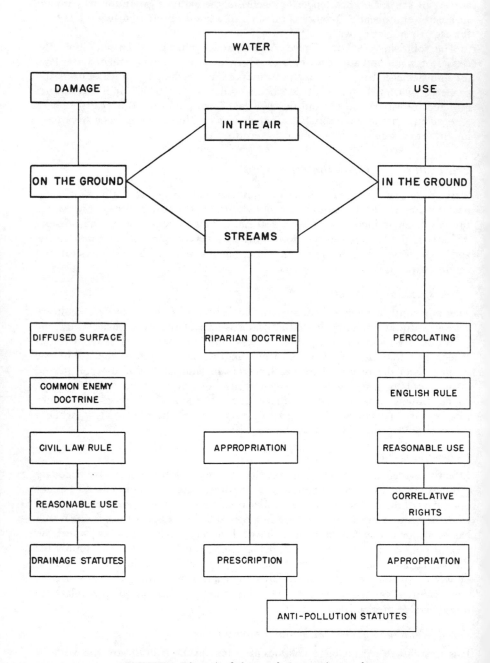

FIGURE 1. Diagram of classes of water and water laws.

to use of stream water in the riparian owners, whether they choose to use the water or not. Efficient water management suggests that rights to stream waters be correlated to proposed uses, rather than to the geographic location of users' lands. A single ground-water source often supplies many users, but under the common law a single landowner may seriously impair this general source without any public opportunity to restore the source or to restore the well yields of others depending upon it. Wherever effective water management becomes urgent, ground water sources should never be impaired.

Waste and Doubtful Water Rights

Water users like to be assured that their investments are secure and that they may safely proceed to use the water. When the user's rights to a stream are doubtful, he is discouraged from investing in new construction and equipment and from proceeding to use the water. This in turn might tend to cause much water to go unused. Under non-statutory law, the user's rights often are not fully determined until a suit has been brought against him and a court has ruled. Thus irrigation, which is rapidly becoming more popular as a high-yield, high-income farming method, is discouraged so long as irrigation farmers realize that downstream riparian owners can bar this consumptive use of stream waters. Plenty of stream water is available for widespread irrigation, but the practice is discouraged by non-statutory law.

An industrial plant may make "reasonable" consumptive use of stream waters, but the meaning of "reasonable" sometimes must await court interpretation, and investment might be discouraged. Still further, an industrial plant might begin operations, then find that a similar plant, subsequently established upstream, may take precedence over the lower plant during times of short supply. Even the surface supplies of public utilities and public waterworks depend in some measure upon the power of eminent domain and upon the payment of riparian owners for the right to use stream waters.

Evaluating Water Rights Law

Despite the defects of non-statutory water rights law generally, it may not yet be concluded that Ohio should abandon or substantially change its present legal structure. It should be noted that each of the criticisms relates to water management. Each of those criticisms becomes more important as the problem of water supply becomes more critical.

For example, when water is ample to meet all requirements there is little need to classify water uses as to the more essential and less essential uses. Both more essential and less essential needs can be met without serious conflicts over water rights. Where there is plenty of water, the arbitrary common-law distinctions between ground and surface sources are not too important, since serious impairment of either source is not likely to occur. Riparian and/or ground-water rights may tend to encourage waste of water, but this waste is not too serious if the supply is more than adequate for all demands in all localities. Riparian or ground-water rights may be unclear, or they may tend to encourage impairment of water sources. But again, these factors are not too significant if every user has plenty of usable water at the time and place of need.

For these reasons an evaluation of Ohio's water rights law must include an evaluation of Ohio's physical and economic problems with respect to water. The kind of law needed depends upon the amount and distribution of usable water in the state. Where the supply is so great and so well distributed that there never is local shortage nor conflict over water rights, there is little concern over the defects of the common law. But where the supply of usable water is less adequate and less well distributed, or promises to be in the future, there is relatively greater need to re-examine and perhaps to modify the existing water rights law.

In order to begin this evaluation, it is necessary to survey the state's water supply, its magnitude and distribution; to examine the trends in demand and discover whether there are increasing physical and economic difficulties in meeting the demand; to study recent problems in water management and discover whether existing water rights law has, in fact, impeded the efforts to provide adequate usable water for all.

If it should be found that Ohio's water supply problems are becoming increasingly severe and that present water rights law is or will be a serious contributor to those problems, then it may be concluded that the law should be modified.

Footnotes

1. East Bay Sporting Club v. Miller, 118 OS 36.
2. Frazier v. Brown, 12 OS 294; Logan Gas Co. v. Glasgo, 122 OS 126.
3. Frazier v. Brown, 12 OS 294.
4. Busby, C. E., Interview, 6 July, 1954. Mr. Busby, a soil conservationist on the immediate planning staff of the U.S. Secretary of Agriculture, is one of the country's outstanding authorities on problems of water law.
5. *Ibid.*
6. Crawford v. Rambo, 44 OS 279.
7. *Ibid.;*
8. Ulland v. Foss, Schneider Brewing Co., 20 ONPNS 375, reversed in 28 OCA 529, which was reversed in 97 OS 210.
9. Logan Gas Co. v. Glasgo, 122 OS 126.
10. Weil, S. C., 1919. Waters: American law and French authority, Harvard Law Review 33:136.
11. Busby, C. E., Water rights and our expanding economy. J. of Soil and Water Conservation 9 (March 1954):68.
12. *Ibid.*, and Weil, *Op. Cit.*
13. Busby, 1954. Water rights and our expanding economy, *Op Cit.*
14. Farnham, H. P., 1904. The law of water and water rights. Rochester: Lawyers' Cooperative Publishing Co., vol I, p. 280ff.
15. *Ibid.*, p. 284, see also 27 Ruling Case Law, Sec. 22, p. 1079.
16. East Bay v. Miller, *Op. Cit.*
17. Crawford v. Rambo, 44 OS 279.
18. City of Canton v. Shock, 66 OS 19.
19. Deming v. Cleveland, 22 OCC 1.
20. Cooper v. Hall, 5 O 320.
21. *Ibid.*
22. *Ibid.*, also Dissette v. Lowerie, 9 (Dec. 545).

23. Canton v. Shock, 66 OS 19.
24. Turner Co. v. Holly Mfg. Co., 12 OCC 738.
25. Canton v. Shock, *Op. Cit.*
26. *Ibid.*
27. Lambeck v. Nye, 47 OS 336.
28. See 41 O. Jur. and the note at 221 in 70 ALR.
29. *Ibid.* and Dayton v. Robert, 8 OCC 64.
30. *Ibid.*
31. Straight v. Hover, 79 OS 263; Mansfield v. Balliet, 65 OS 451; Kirk v. Cincinnati, 25 ONPS 473.
32. O'Neil, C. W., 1954. Water rights and the legal aspects of water diversion between watersheds. Address to Water Clinic Conference at Columbus, Ohio, 11 February 1954.
33. Busby, 1954. Water rights and our expanding economy, *Op. Cit.*
34. Report of the President's Water Resources Policy Commission, Vol. III, Water Resources Law, p. 156ff.
35. Arizona Constitution, Article 17, Sec. 1.
36. Arizona Code: Waters.
37. Colorado Constitution, Article 15, Sec. 5.
38. *Ibid.*, Sec. 6.
39. Colorado Statutes Annotated, Ch. 90
40. Idaho Constitution, Article 15, Sec. 2.
41. Idaho Code, Sec. 42.
42. Several Idaho cases support this statement: Crane Falls Power & Irrigation Co. v. Snake River Irrigation Co., 24 Idaho 63; Reno v. Richards, 32 Idaho 1; Silkey v. Tiegs, 51 Idaho 344; Bachman v. Reynolds Irrigation District, 56 Idaho 344.
43. State ex. rel. Peterson v. State Board of Agriculture, 158 Kansas 603.
44. State ex. rel Emery v. Knapp, 167 Kansas 546.
45. Kansas Laws, Chapter 390.
46. Noe, W. L. 1954. Kansas' experience in water rights legislation, Unpublished address delivered before Mid-Western States Flood Control Conference, East Lansing, Michigan, 16 June, 1954. Mr. Noe is a special assistant attorney general of Kansas.
47. *Ibid.*
48. See Report of President's Water Policy Commission, Vol. III. Appendix B, p. 734ff, and Montana Revised Code, Sec. 89.
49. Montana Revised Code, Sec. 89.
50. Nevada General Laws Compiled, 1929, and Supplements, Secs. 7890-8254.
51. New Mexico Revised Statutes, 1941, and 1951 Supplement, Chapter 77.
52. 7 Utah Code Annotated, title 73.
53. Wyoming Compiled Statutes, title 71.
54. Cases bearing are Lux v. Haggin, 69 California 255; Hermershaus v. Southern California Edison Co., 200 California 81; and Peabody v. Vallejo, 2 California (2nd) 351. For more information and discussion of this issue see Deering's California Water Code, 1949, and title Waters in Cal. Jour. and Cal. Digest.
55. Crawford Company v. Hathaway, 67 Nebraska 325; Ming v. Coffee, 67 Nebraska 500.
56. McCook Irrigation and Water Power Co. v. Guthrie, 70 Nebraska 109; Cline v. Stock, 71 Nebraska 70.
57. See Revised Statutes of Nebraska, Chapter 46.
58. For further information see Report of the President's Water Policy Commission, Vol. III, Water Resources Law, p. 748-49; North Dakota Constitution and 5 North Dakota Revised Code, title 61, Waters.
59. Report of the President's Water Policy Commission, Vol. II, Appendix B, p. 756.
60. Water Rights of Hood River, 114 Oregon 112; and California-Oregon Power Co. v. Beaver Portland Cement Co., 73 Fed (2nd) 555.
61. Section 8, Oregon Compiled Laws Annotated, Water Code, Secs. 116-1206.
62. Rhodes v. Whitehead, 27 Texas 304 (1836); and Chicago R. I. & Gulf Ry. v. Tarrant County, W. C. & I. District No. 1, 123 Texas 432 (1934). See also 21 Texas Civil

      Statutes (1925), and Report of the President's Water Policy Commission, Vol. III,
      Appendix B.
63. 6 Revised Code, Washington, Chapter 90.
64. Busby, C. E. American Water Rights Law, South Carolina Law Quarterly 5:111.
65. Frazier v. Brown, 12 OS 303.
66. *Ibid.*, and Wyandot Club v. Sells, 60 ONP 64.
67. Logan Gas Co. v. Glasgo, 122 OS 126.
68. Frazier v. Brown, 12 OS 294; Castalia Trout Club v. Castalia Sporting Club, 8 OCC
      194; Warder v. Springfield, 90 Dec. Rep. 855; and also see the note in ALR 55:1397.
69. This interpretation, based on writer's investigation, concurs with the interpretation in
      Harry B. Reese's unpublished manuscript on Ohio's water laws; this project is a
      cooperative study sponsored by the Ohio State University College of Law and the
      Division of Water.
70. Ohio Colleries v. Cocke; and Natural Coal Co. v. Gaffer, 107 OS 238.
71. Bassett v. Osborn, 23 OCCNS 342.
72. Katz v. Walkinshaw, 141 California 116.

Enforcement of 1899 Refuse Act Through Citizen Action

I. What is Prohibited and Where

The 1899 Refuse Act is a powerful, but little used, weapon in our Federal arsenal of water pollution control enforcement legislation. Section 13 of the Act (Title 33, United States Code, section 407) prohibits anyone, including any individual, corporation, municipality, or group, from throwing, discharging, or depositing any refuse matter of any kind or any type from a vessel or from a shore-based building, structure, or facility into either (a) the Nation's navigable rivers, lakes, streams, and other navigable bodies of water, or (b) any tributary to such waters, unless he has first obtained a permit to do so. Navigable water includes water sufficient to float a boat or log at high water. This section of the Act applies to inland waters, coastal waters, and waters that flow across the boundaries of the United States and Canada and Mexico.

The term "refuse" has been broadly defined by the Supreme Court to include all foreign substances and pollutants. It includes solids, chemicals, oils, and other liquid pollutants. The only materials excepted from this general prohibition are those flowing from streets, such as from storm sewers, and from municipal sewers, which pass into the waterway in liquid form.

In addition, the section prohibits anyone from placing on the bank of any navigable waterway, or of any tributary to such waterway, any material that could be washed into a waterway by ordinary or high water, or by storms or floods, or otherwise and would result in the obstruction of navigation.

II. Permits to Discharge

Section 13 of the Act authorizes the Secretary of the Army, acting through the Corps of Engineers, to permit the deposit of material into navigable waters under conditions prescribed by him. Regulations governing the issuance of permits are published in Title 33 of the Code of Federal Regulations, Part 209.

Reprinted from an edited and unpublished report by the Conservation and Natural Resources Subcommittee of the Committee on Government Operations, 91st Cong. 2d sess., 1970.

III. Penalty for Violations

Violations of the Refuse Act are subject to criminal prosecution and penalties of a fine of not more than $2500 nor less than $500 for each day or instance of violation, or imprisonment for not less than 30 days nor more than 1 year, or both a fine and imprisonment (Title 33, United States Code, Section 411). A citizen, who informs the appropriate United States attorney about a violation and gives sufficient information to lead to conviction, is entitled to one-half of the fine set by the court. (See section V of this outline.)

IV. Procedure for Citizen to Seek Enforcement of Refuse Act

A. The citizen having information about any discharge of refuse into navigable waters should first *ascertain whether the discharge is authorized by Corps permit.* If a permit is in effect, the citizen should endeavor to ascertain whether the *permittee is complying with its terms.* This information can be obtained from the appropriate office of the Corps of Engineers with jurisdiction over the particular waters into which the discharge occurs. Such information is available to the public under the Freedom of Information Act (5 U.S. Code 552; Public Law 90-23).

B. The Refuse Act specifically directs that the appropriate *United States attorney* shall *"vigorously prosecute all offenders."* (Title 33, United States Code, section 413.) In order to do so he needs adequate information to prove that the discharges were made and that they violated the law or the conditions of the permit. Furthermore, the statute specifies that the citizen's right to one-half of the fine is conditioned on his providing to the U.S. attorney information sufficient to *lead to a conviction of the violator.*

In providing information to the U.S. Attorney, the citizen should make a detailed statement, sworn to before a notary or other officer authorized to administer oaths, setting forth:

(a) the nature of the refuse material discharged;

(b) the source and method of discharge;

(c) the location, name, and address of the person or persons causing or contributing to the discharge;

(d) the name of the waterway into which the discharge occurred;

(e) each date on which the discharge occurred;

(f) the names and addresses of all persons known to you, including yourself, who saw or knows about the discharges and could testify about them if necessary;

(g) a statement that the discharge is not authorized by Corps permit, or, if a permit was granted, state facts showing that the alleged violator is not complying with any condition of the permit;

(h) if the waterway into which the discharge occurred is not commonly known as navigable, or as a tributary to a navigable waterway, state facts to show such status;

(i) Where possible, photographs should be taken, and samples of the pollutant or foreign substance collected in a clean jar which is then sealed. These should be labeled with information showing who took the photograph or sample, where, and when, and how; and who retained custody of the film or jar.

Where the material is liable to be washed into the waterway from its bank, in violation of the Act, similar information should also be provided to the U.S. attorney in such a statement.

C. When a citizen furnishes information to the U.S. attorney for the purpose of aiding in the prosecution of violators of the Refuse Act for past discharges, the citizen should also urge the U.S. attorney *to seek injunctions* under the same Act to *preclude future discharges,* or other orders to require the dischargers to remove pollutants already discharged. More frequent use of this authority by the Government, together with criminal sanctions, will have lasting pollution control results.

V. *"Qui Tam" Suits*

Where a statute, such as the Refuse Act, provides that part of a fine shall be paid to citizens who furnish sufficient information of a violation to lead to a conviction of the violator, and the *Government fails to prosecute within a reasonable period of time, the informer can bring his own suit,* in the name of the Government, against the violator to collect his portion of the penalty. This is called a *"qui tam"* suit. The informer has a financial interest in the fine and therefore can sue to collect it. The Supreme Court has upheld such *qui tam* suits. Some of these decisions are cited in the Report of the House Committee on Government Operations (House Report 91-917, March 18, 1970) entitled "Our Waters and Wetlands: How the Corps of Engineers can help prevent their Destruction and Pollution."

The *United States district courts* apparently have exclusive jurisdiction to hear and decide such suits. (Title 28, United States Code, section 1355.) In such a *qui tam* suit, the citizen must prove that the alleged violator did, in fact, violate the Act.

If the citizen should lose the suit, he probably would have to bear the cost of suing, including his lawyer's fees.

The Refuse Act consists of various provisions of the River and Harbor Act of 1899 (30 Stat. 1151) and other statutes relating to the discharge of refuse into navigable waters, as follows:

Sec. 407—Deposit of refuse in navigable waters generally.

It shall not be lawful to throw, discharge, or deposit, or cause, suffer, or procure to be thrown, discharged, or deposited either from or out of any ship, barge, or other floating craft of any kind, or from the shore, wharf, manufacturing establishment, or mill of any kind, any refuse matter of any kind or description whatever other than that flowing from streets and sewers and passing therefrom in a liquid state, into any navigable water of the United States, or into any tributary of any navigable water from which the same shall float or be washed into such navigable water; and it shall not be lawful to deposit, or cause, suffer, or procure to be deposited material of any kind in any place on the bank of any navigable water, or on the bank of any tributary of any navigable water, where the same shall be liable to be washed into such navigable water, either by ordinary or high tides, or by storms or floods, or otherwise, whereby navigation shall or may be impeded or obstructed: *Provided,* That nothing herein contained shall extend to, apply to, or prohibit the operations in connection with the improvement of navigable waters or construction of public works, considered necessary and proper by the United States officers supervising such improvement or public work: *And provided further,* That the Secretary of the Army, whenever in the judgment of the Chief of Engineers anchorage and navigation will not be injured thereby, may permit the deposit of any material above mentioned in navigable waters, within limits to be defined and under conditions to be prescribed by him, provided application is made to him prior to depositing such material; and whenever any permit is so granted the conditions thereof shall be strictly complied with, and any violation thereof shall be unlawful. (Mar. 3, 1899, ch. 425, Sec. 13, 30 Stat. 1152.)

●   ●   ●

Sec. 411—Penalty for wrongful deposit of refuse; use of or injury to harbor improvements, and obstruction of navigable waters generally.

Every person and every corporation that shall violate, or that shall knowingly aid, abet, authorize, or instigate a violation of the provisions of sections 407, 408, and 409 of this title shall be guilty of a misdemeanor, and on conviction thereof shall be punished by a fine not exceeding $2,500 nor less than $500, or by imprisonment (in the case of a natural person) for not less than thirty days nor more than one year, or by both such fine and imprisonment, in the discretion of the court, one-half of said fine to be paid to the person or persons giving information which shall lead to conviction. (Mar. 3, 1899, ch. 425, Sec. 16,   30 Stat. 1153).

Sec. 412—Liability of masters, pilots, and so forth, and of vessels engaged in violations.

Any and every master, pilot, and engineer, or person or persons acting in such capacity, respectively, on board of any boat or vessel who shall knowingly engage in towing any scow, boat, or vessel loaded with any material specified in section 407 of this title to any point or place of deposit or discharge in any harbor or navigable water, elsewhere than within the limits defined and permitted by the Secretary of the Army, or who shall willfully injure or destroy any work of the United States contemplated in section 408 of this title, or who shall willfully obstruct the channel of any waterway in the manner contemplated in section 409 of this title, shall be deemed guilty of a violation of sections 401, 403, 404, 406, 407, 408, 409, 411 - 416, 418, 502, 549, 686, and 687 of this title, and shall upon conviction be punished as provided in section 411 of this title, and shall also have his license revoked or suspended for a term to be fixed by the judge before whom tried and convicted. And any boat, vessel, scow, raft, or other craft used or employed in violating any of the provisions of sections 407, 408, and 409 of this title shall be liable for the pecuniary penalties specified in section 411 of this title, and in addition thereto for the amount of the damages done by said boat, vessel, scow, raft, or other craft, which latter sum shall be placed to the credit of the appropriation for the improvement of the harbor or waterway in which the damage occurred, and said boat, vessel, scow, raft, or other craft may be proceeded against summarily by way of libel in any district court of the United States having jurisdiction thereof. (Mar. 3, 1899, ch. 425, Sec. 16, 30 Stat. 1153.)

Sec. 413—Duty of United States attorneys and other Federal officers in enforcement of provisions; arrest of offenders.

The Department of Justice shall conduct the legal proceedings necessary to enforce the provisions of sections xxx 407, 408, 409,[and] 411, xxx of this title; and it shall be the duty of United States attorneys to vigorously prosecute all offenders against the same whenever requested to do so by the Army or by any of the officials hereinafter designated, and it shall furthermore be the duty of said United States attorneys to report to the Attorney General of the United States the action taken by him against offenders so reported, and a transcript of such reports shall be transmitted to the Secretary of the Army by the Attorney General; and for the better enforcement of the said provisions and to facilitate the detection and bringing to punishment of such offenders, the officers and agents of the United States in charge of river and harbor improvements, and the assistant engineers and inspectors employed under them by authority of the Secretary of the Army, and the United States collectors of customs and other revenue officers shall have power and authority to swear out process, and to arrest and take into custody, with or without process, any person or persons who may commit any of the acts or offenses prohibited by the said sections, or who may violate any of the provisions of the same: *Provided,* That no person shall be arrested without process for any offense not committed in the presence of some one of the aforesaid officials: *And provided further,* That whenever any arrest is made under such sections, the person so arrested shall be brought forthwith before a commissioner, judge, or court of the United States for examination of the offenses alleged against him; and such commissioner, judge, or court shall proceed in respect thereto as authorized by law in case of crimes against the United States. (Mar. 3, 1899, ch. 425, Sec. 17, 30 Stat. 1153; June 25, 1948, ch. 646, Sec. 1, 62 Stat. 909, eff. Sept. 1, 1948.)

Sec. 417—Expenses of investigations by Department of the Army.

Expenses incurred by the Engineer Department of the Department of the Army in all investigations, inspections, hearings, reports, service of notice, or other action incidental to examination of plans or sites of bridges or other structures built or proposed to be built in or over navigable waters, or to examinations into alleged violations of laws for the protection and preservation of navigable waters, or to the establishment or marking of harbor lines, shall be payable from any funds which may be available for the improvement, maintenance, operation, or care of the waterways or harbors affected, or if such funds are not available in sums judged by the Chief of Engineers to be adequate, then from any funds available for examinations, surveys, and contingencies of rivers and harbors. (Mar. 3, 1905, ch. 1482, Sec. 6, 33 Stat. 1148.)

Sec. 419—Regulation by Secretary governing transportation and dumping of dredgings, refuse, etc., into navigable waters; oyster lands; appropriations.

The Secretary of the Army is authorized and empowered to prescribe regulations to govern the transportation and dumping into any navigable water, or waters adjacent thereto, of dredgings, earth, garbage, and other refuse materials of every kind or description, whenever in his judgment such regulations are required in the interest of navigation. Such regulations shall be posted in conspicuous and appropriate places for the information of the public; and every person or corporation which shall violate the said regulations, or any of them, shall be deemed guilty of a misdemeanor and shall be subject to the penalties prescribed in sections 411 of this title, for violation of the provisions of section 407 of this title: *Provided,* That any regulations made in pursuance hereof may be enforced as provided in section 413 of this title, the provisions whereof are made applicable to the said regulations: *Provided further,* That this section shall not apply to any waters within the jurisdictional boundaries of any State which are now or may hereafter be used for the cultivation of oysters under the laws of such State, except navigable channels which have been or may hereafter be improved by the United States, or to be designated as navigable channels by competent authority, and in making such improvements of channels, the material dredged shall not be deposited upon any ground in use in accordance with the laws of such State for the cultivation of oysters, except in compliance with said laws: *And provided further,* That any expense necessary in executing this section may be paid from funds available for the improvement of the harbor or waterway, for which regulations may be prescribed, and in case no such funds are available the said expense may be paid from appropriations made by Congress for examinations, surveys, and contingencies of rivers and harbors. (Mar. 3, 1905, ch. 1482, Sec. 4, 33 Stat. 1147.)

Sec. 421—Deposit of refuse, etc., in Lake Michigan near Chicago.

It shall not be lawful to throw, discharge, dump, or deposit, or cause, suffer, or procure, to be thrown, discharged, dumped, or deposited, any refuse matter of any kind or description whatever other than that flowing from streets and sewers and passing therefrom in a liquid state into Lake Michigan, at any point opposite or in front of the county of Cook, in the State of Illinois, or the county of Lake in the State of Indiana, within eight miles from the shore of said lake, unless said material shall be placed inside of a breakwater so arranged as not to permit the escape of such refuse material into the body of the lake and cause contamination thereof; and no officer of the Government shall dump or cause or authorize to be dumped any material contrary to the provisions of this section: *Provided, however,* That the provisions of this section shall not apply to work in connection with the construction, repair, and protection of breakwaters and other structures built in aid of navigation, or for the purpose of obtaining water supply. Any person violating any provision of this section shall be guilty of a misdemeanor, and on conviction thereof shall be fined for each offense not exceeding $1000. (June 23, 1910, ch. 359, 36 Stat. 593.)

New York Harbor, Harbor of Hampton Roads, and Harbor of Baltimore

Sec. 441—Deposit of refuse prohibited; penalty.

The placing, discharging, or depositing, by any process or in any manner, of refuse, dirt, ashes, cinders, mud, sand, dredgings, sludge, acid, or any other matter of any kind, other than that flowing from streets, sewers, and passing therefrom in a liquid state, in the waters of any harbor subject to sections 441 - 451b of this title, within the limits which shall be prescribed by the supervisor of the harbor, is strictly forbidden, and every such act is made a misdemeanor, and every person engaged in or who shall aid, abet, authorize, or instigate a violation of this section, shall, upon conviction, be punishable by fine or imprisonment, or both, such fine to be not less than $250 nor more than $2,500, and the imprisonment to be not less than thirty days nor more than one year, either or both united, as the judge before whom conviction is obtained shall decide, one-half of said fine to be paid to the person or persons giving information which shall lead to conviction of this misdemeanor. (June 29, 1888, ch. 496, Sec. 1, 25 Stat. 209; Aug. 28, 1958, Pub. L. 85-802, Sec. 1 (1), 72 Stat. 970.)

Sec. 442—Liability of officers of towing vessel.

Any and every master and engineer, or person or persons acting in such capacity, respectively, on board of any boat or vessel, who shall knowingly engage in towing any scow, boat, or vessel loaded with any such prohibited matter to any point or place of deposit, or discharge in the waters of any harbor subject to sections 441 - 451b of this title, or to any point or place elsewhere than within the limits defined and permitted by the supervisor of the harbor, shall be deemed guilty of a violation of section 441 of this title, and shall upon conviction, be punishable as provided for offenses in violation of section 441 of this title, and shall also have his license revoked or suspended for a term to be fixed by the judge before whom tried and convicted. (June 29, 1888, ch. 496, Sec. 2, 25, Stat. 209; Aug. 28, 1958, Pub. L. 85-802, Sec. 1 (2), 72 Stat. 970.)

Sec. 443—Permit for dumping; penalty for taking or towing boat or scow without permit.

In all cases of receiving on board of any scows or boats such forbidden matter or substance as described in section 441 of this title, the owner or master, or person acting in such capacity on board of such scows or boats, before proceeding to take or tow the same to the place of deposit, shall apply for and obtain from the supervisor of the harbor appointed, as provided in section 451 of this title, a permit defining the precise limits within which the discharge of such scows or boats may be made; and it shall not be lawful for the owner or master, or person acting in such capacity, of any tug or towboat to tow or move any scow or boat so loaded with such forbidden matter until such permit shall have been obtained; and every person violating the foregoing provisions of this section shall be deemed guilty of a misdemeanor, and on conviction thereof shall be punished by a fine of not more than $1,000 nor less than $500, and in addition thereto the master of any tug or towboat so offending shall have his license revoked or suspended for a term to be fixed by the judge before whom tried and convicted. (June 29, 1888, ch. 496, Sec. 3, 25 Stat. 209; Aug. 18, 1894, ch. 299, Sec. 3, 28 Stat. 360; May 28, 1908, ch. 212, Sec. 8, 35 Stat. 426.)

Sec. 444—Dumping at other place than designated dumping grounds; penalty; person liable; excuses for deviation.

Any deviation from such dumping or discharging place specified in such permit shall be a misdemeanor, and the owner and master, or person acting in the capacity of master, of any scows or boats dumping or discharging such forbidden matter in any place other than that specified in such permit shall be liable to punishment therefor as provided in section 441 of this title; and the owner and master, or person acting in the capacity of master, of any tug or towboat towing such scows or boats shall be liable to equal punishment with the owner and master, or person acting in the capacity of master, of the scows or boats; and further, every scowman or other employee on board of both scows and towboats shall be deemed to have knowledge of the place of dumping specified in such permit, and the owners and masters, or persons acting in the capacity of masters, shall be liable to punishment, as aforesaid, for any unlawful dumping, within the meaning of sections 441 - 452 of this title, which may be caused by the negligence of ignorance of such scowman or other employee; and further, neither defect in machinery nor avoidable accidents to scows or towboats, nor unfavorable weather, nor improper handling or moving of scows or boats of any kind whatsoever shall operate to release the owners and master and employees of scows and towboats from the penalties mentioned in section 441 of this title. (June 29, 1888, ch. 496, Sec. 3, 25 Stat. 209; Aug. 18, 1894, ch. 299, Sec. 3, 28 Stat. 360; May 28, 1908, ch. 212, Sec. 8, 35 Stat. 426.)

Sec. 449 —Disposition of dredged matter; persons liable; penalty.

All mud, dirt, sand, dredgings, and material of every kind and description whatever taken, dredged, or excavated from any slip, basin, or shoal in any harbor subject to sections 441 - 451b of this title, and placed on any boat, scow, or vessel for the purpose of being taken or towed upon the waters of that harbor to a place of deposit, shall be deposited and discharged at such place or within such limits as shall be defined and specified by the supervisor of the harbor, as in section 443 of this title prescribed, and not otherwise. Every person, firm, or corporation being the owner of any slip, basin, or shoal, from which such mud, dirt, sand, dredgings, and material shall be taken, dredged, or excavated, and every person, firm, or corporation in any manner engaged in the work of dredging or excavating any such slip, basin, or shoal, or of removing such mud, dirt, sand, or dredgings therefrom, shall severally be responsible for the deposit and discharge of all such mud, dirt, sand, or dredgings at such place or within such limits so defined and prescribed by said supervisor of the harbor; and for every violation of the provisions of this section the person offending shall be guilty of an offense, and shall be punished by a fine equal to the sum of $5 for every cubic yard of mud, dirt, sand, dredgings, or material not deposited or discharged as required by this section. (June 29, 1888, ch. 496, Sec. 4, 25 Stat. 210; Aug. 28, 1958, Pub. L 85-802, Sec. 1 (5), 72 Stat. 970.)

Sec. 450—Liability of vessel.

Any boat or vessel used or employed in violating any provision of sections 441 - 451b of this title, shall be liable to the pecuniary penalties imposed thereby, and may be proceeded against, summarily by way of libel in any district court of the United States having jurisdiction thereof. (June 29, 1888, ch. 496, Sec. 4, 25 Stat. 210.)

Sec. 451a.—Harbors subject to sections 441 - 451b of this title.

The following harbors shall be subject to sections 441 - 451b of this title:

(1) The harbor of New York.
(2) The harbor of Hampton Roads.
(3) The harbor of Baltimore.

(June 29, 1888, ch. 496, Sec. 6, 25 Stat. 210; Aug. 28, 1958, Pub. L. 85-802, Sec. 1 (7), 72 Stat. 970.)

Sec. 451b.—Same; waters included.

For the purposes of sections 441 - 451b of this title—

(1) The term "harbor of New York" means the tidal waters of the harbor of New York, its adjacent and tributary waters, and those of Long Island Sound.

(2) The term "harbor of Hampton Roads" means the tidal waters of the harbors of Norfolk, Portsmouth, Newport News, Hampton Roads, and their adjacent and tributary waters, so much of the Chesapeake Bay and its tributaries as lies within the State of Virginia, and so much of the Atlantic Ocean and its tributaries as lies within the jurisdiction of the United States within or to the east of the State of Virginia.

(3) The term "Harbor of Baltimore" means the tidal waters of the harbor of Baltimore and its adjacent and tributary waters, and so much of Chesapeake Bay and its tributaries as lie within the State of Maryland. (June 29, 1888, ch. 496, Sec. 7, as added Aug. 28, 1958, Pub. L 85-802, Sec. 1 (8), 72 Stat. 970.)

Our Waters and Wetlands:
How the Corps of Engineers Can Help
Prevent Their Destruction and Pollution

On March 17, 1970, the Committee on Government Operations approved and adopted a report entitled "Our Waters and Wetlands: How the Corps of Engineers Can Help Prevent Their Destruction and Pollution." The chairman was directed to transmit a copy to the Speaker of the House.

The natural environments of our Nation's bays, estuaries, and other water bodies are being destroyed or threatened with destruction by water pollution, alteration of river courses, landfilling of the shallow and marshland areas, sedimentation, dredging, construction of piers and bulkheads, and other manmade changes. Many of these water areas, including some located near densely populated urban areas, serve public needs for recreational opportunities and provide feeding, habitat, and nesting or spawning grounds for migratory waterfowl, fish, shellfish, and other wildlife. Many Federal agencies participate in, or authorize work and activities which contribute to, the destruction of these water areas, and some agencies have specific responsibilities for preventing such pollution and destruction.

This report examines several aspects of the Corps of Engineers' role in carrying out its responsibilities for protecting the Nation's water areas, and recommends how the Corps can stop or minimize this pollution and destruction.

I. THE CORPS OF ENGINEERS, WHICH IS CHARGED BY CONGRESS WITH THE DUTY TO PROTECT THE NATION'S NAVIGABLE WATERS, SHOULD, WHEN CONSIDERING WHETHER TO APPROVE APPLICATIONS FOR LANDFILLS, DREDGING AND OTHER WORK IN NAVIGABLE WATERS, INCREASE ITS CONSIDERATION OF THE EFFECTS WHICH THE PROPOSED WORK WILL HAVE, NOT ONLY ON NAVIGATION, BUT ALSO ON CONSERVATION OF NATURAL

Reprint of a report by the Committee on Government Operations, 91st Cong., 2d sess., 1970, H. Rept. 91-917, pp. 1-18.

RESOURCES, FISH AND WILDLIFE, AIR AND WATER QUALITY, ESTHETICS, SCENIC VIEW, HISTORIC SITES, ECOLOGY, AND OTHER PUBLIC INTEREST ASPECTS OF THE WATERWAY

The River and Harbor Act of 1899 (act of March 3, 1899, c. 425, 30 Stat. 1151), authorizes the Corps of Engineers to regulate or prevent the filling of land submerged by water at high tide. That act forbids the creation of "any obstruction not affirmatively authorized by Congress to the navigable capacity of any of the waters of the United States" (sec. 10; 33 U.S.C. sec. 403), or the discharge or deposit in navigable water or its tributaries of "any refuse matter of any kind or description whatever other than that flowing from streets and sewers and passing therefrom in a liquid state" (sec. 13; 33 U.S.C. 407), unless such work, discharge or deposit is done under a permit from the Corps of Engineers approved by the Secretary of the Army.

For many years, the Corps of Engineers administered the River and Harbor Act of 1899 with primary or exclusive emphasis on how the proposed structure or fill would affect navigation. Indeed, until about 2 years ago, the Corps' public notices announcing the filing of applications for permits to fill, dredge or construct works in navigable waters, defined the Corps' interest as being confined to issues of navigation, and requested comments from the public only on such issues.

That restricted view of the 1899 act, however, was not required by the law. As early as 1933 the Supreme Court ruled that under the 1899 act the Department of the Army may properly deny a permit to erect a structure (or to make a fill) in the Potomac River, if such permit would interfere with the public interest in having a parkway or recreation area. *United States ex rel. Greathouse v. Dern*, 289 U.S. 352 (1933).

The scope of the 1899 act was further amplified by the Fish and Wildlife Coordination Act (as amended by the act of August 12, 1958, 72 Stat. 563, Public Law 85-624; 16 U.S.C. 661). This act enunciates a national policy of "recognizing the vital contribution of our wildlife resources to the Nation, the increasing public interest and significance thereof due to expansion of our national economy and other factors, and to provide that wildlife conservation shall receive equal consideration and be coordinated with other features of water-resource development programs. . . ." (Sec. 1; 16 U.S.C. 661.)

To carry out that policy, section 2(a) of the act (16 U.S.C. 662(a)) directs that "whenever the waters of any stream or other body of water are proposed . . . to be impounded, diverted, the channel deepened, or . . . otherwise controlled or modified for any purpose whatever . . . by any public or private agency under Federal permit or license . . . such . . . agency first shall consult with the United States Fish and Wildlife Service, Department of the Interior . . . with a view to the conservation of wildlife resources by preventing loss or damage to such resources . . . ."

Subsection 2(b) of the act further provides that the "reports and recommendations of the Secretary of the Interior on the wildlife aspects of such projects . . . based on surveys and investigations conducted by the United States Fish and Wildlife Service . . . for the purpose of determining means and measures that should be adopted to prevent the loss of or damage to such wildlife resources, as well as to provide concurrently for the development and improvement of such resources, shall be made an integral part of any report prepared or submitted . . . to any agency . . . having the authority . . . to authorize the construction of water-resource development projects. . . . Recommendations of the Secretary of the Interior shall be as specific as is practicable with respect to features recommended for wildlife conservation and development, . . . the results expected, and shall describe the damage to wildlife attributable to the project and the measures proposed for mitigating or compensating for these damages." The section further requires that "full consideration" must be given to the Secretary's report and recommendations "on the wildlife aspects."

The Corps' obligation to consider all facets of the public interest in protecting estuaries, rivers, lakes, and other navigable waters also arises from the national policy and directives expressed in many other statutes and Executive orders designed to minimize pollution, maximize recreation, protect esthetics, preserve natural resources, and promote the comprehensive planning and use of water bodies to enhance the public interest rather than private gain.[1]

[1] Some of these statutes and orders are:

Federal Water Power Act of 1920, as amended (16 U.S.C. 791, et seq.);

Oil Pollution Act of 1924, as amended (33 U.S.C. 431, et seq.);

Federal Water Pollution Control Act (33 U.S.C. 466, et seq.) as amended by Water Quality Act of 1965 (Public Law 89-234; 79 Stat. 903) and Clean Water Restoration Act of 1966 (Public Law 89-753; 80 Stat. 1246);

Delaware River Basin Compact Act of September 27, 1961 (Public Law 87-328; 75 Stat. 688);

Bureau of Outdoor Recreation Act of May 28, 1963 (Public Law 88-29, 77 Stat. 49; 16 U.S.C. 460 1);

Section 212, Appalachian Regional Development Act of March 9, 1965 (79 Stat. 5, 16; Public Law 89-4; 40 U.S.C. App. sec. 212);

Water Resources Planning Act of July 22, 1965 (Public Law 89-80; 79 Stat. 244; 42 U.S.C. 1962a);

Section 702, Housing and Urban Development Act of August 10, 1965 (79 Stat. 451, 490; Public Law 89-117; 42 U.S.C. 3102);

Section 106, Public works and Economic Development Act of August 26, 1965 (79 Stat. 552, 554; 42 U.S.C. 3136; Public Law 89-136);

Estuarine Study Act of August 3, 1968 (Public Law 90-454; 82 Stat. 625; 16 U.S.C. 1221);

National Water Commission Act of September 26, 1968 (Public Law 90-515; 82 Stat. 868; 42 U.S.C. 1962a, note);

Executive Order 11288 of July 2, 1966 (3 C.F.R. 423); superseded by Executive Order 11507 of February 4, 1970 (35 F.R. 2573), (prescribing requirements for control and abatement of air and water pollution at Federal installations).

Executive Order 11472 of May 29, 1969 (34 F.R. 8693), as amended by Executive Order 11514 of March 5, 1970 (35 F.R. 4247), (prescribing responsibilities of Federal agencies and the Council on Environmental Quality under the National Environmental Policy Act of 1969, supra).

This national policy is emphasized and more fully expounded in the National Environmental Policy Act of 1969 (Public Law 91-190, 83 Stat. 852), signed by the President on January 1, 1970. In section 102 of the act, Congress mandates (1) that "the policies, regulations, and public laws of the United States shall be interpreted and administered in accordance with the policies" of the National Environmental Policy Act, and (2) that "all agencies of the federal government shall" develop procedures which will "insure that presently unquantified environmental amenities and values" be given "appropriate consideration in decisionmaking along with economic and technical considerations." Section 102 also requires "all agencies of the Federal Government" to prepare a "detailed statement" to be included in "every recommendation or report" concerning "Federal actions significantly affecting the quality of the human environment." That detailed statement must include each of the following matters:

(i) The environmental impact of the proposed action;

(ii) Any adverse environmental effects which cannot be avoided should the proposal be implemented;

(iii) Alternatives to the proposed action;

(iv) The relationship between local short-term uses of man's environment and the maintenance and enhancement of long-term productivity; and

(v). Any irreversible and irretrievable commitments of resources which would be involved in the proposed action should it be implemented.

The act, following the precedent of the consultation requirement in the Fish and Wildlife Coordination Act, also requires that the Federal official making the detailed statement must first "consult with and obtain the comments of any Federal agency which has jurisdiction by law or special expertise with respect to any environmental impact involved." Copies of such statement and comments "shall accompany the proposal through the existing agency review processes." (Sec. 102.)

The combined effect of the 1899 statute, the Coordination Act, the National Environmental Policy Act of 1969, and the many other statutes and executive orders concerning the protection of our waters and other natural resources, is to charge the Corps of Engineers with a responsibility that is analogous to that of the Federal Power Commission in considering an application for a hydropower license. In such cases, the Commission "must include as a basic concern the preservation of natural beauty and of national historic shrines, keeping in mind that, in our affluent society, the cost of a project is only one of several factors to be considered." *Scenic Hudson Preservation Conference v. Federal Power Commission,* 354 F. 2d 608, 624 (C.A. 2, 1965), cert. den., 384 U.S. 941 (1966).

This broad scope of the Corps' duty arises from its power to permit the alteration or destruction of a navigable river or waterway (which Justice Holmes memorably described as "more than an amenity, it is a treasure").[2] The breadth of that duty was further emphasized in 1967 by the Supreme Court of the United States in *Udall v. Federal Power Commission,* 387 U.S. 428 (1967). In

[2]New Jersey v. New York, 283 U.S. 336, 342 (1931).

that case, the Court, in reversing the Commission's grant of a hydropower license, stated (at p. 450):

> The question whether the proponents of a project "will be able to use" the power supplied is relevant to the issue of the public interest. So, too, is the regional need for additional power. But the inquiry should not stop there. A license under the Act empowers the licensee to construct, for its own use and benefit, hydroelectric projects utilizing the flow of navigable waters and thus, in effect, to appropriate water resources from the public domain. The grant of authority to the Commission to alienate Federal water resources does not, of course, turn simply on whether the project will be beneficial to the licensee. Nor is the test solely whether the region will be able to use the additional power. The test is whether the project will be in the public interest. And that determination can be made only after an exploration of all issues relevant to the "public interest," including future power demand and supply, alternate sources of power, the public interest in preserving reaches of wild rivers and wilderness areas, the preservation of anadromous fish for commercial and recreational purposes, and the protection of wildlife.

It is clear that the 1899 act must be read, as the Supreme Court of the United States said in *United States v. Republic Steel Corp.*, 362 U.S. 482, 491 (1960), "charitably in light of the purpose to be served" and not with "a narrow, cramped reading."

Increasing public awareness of the devastation being wrought through indiscriminate filling, dredging and other work in our Nation's waterways has caused the Corps to reexamine and give greater recognition to its responsibilities for the protection of bays, estuaries, and other waterways and to acknowledge that its duty goes beyond simply the question of whether a proposed fill or other work will adversely affect navigation.

On July 13, 1967, the Secretary of the Army and the Secretary of the Interior entered into a memorandum of understanding, establishing procedures whereby the Corps would obtain advice from the Interior Department concerning all effects on fish and wildlife, recreation, pollution, natural resources, or the environment which may arise from dredging, filling or other work authorized by the Corps under the 1899 act.[3]

In 1968, the Corps revised its regulations to state that the Corps, in considering an application for a permit to fill, dredge, discharge or deposit materials, or conduct other activities affecting navigable waters, will evaluate "all relevant factors, including the effect of the proposed work on navigation, fish and wildlife, conservation, pollution, esthetics, ecology, and the general public interest," 33 CFR 209.120(d) (1).[4] The Corps applied this policy when it

[3]The Memorandum of Understanding is reprinted in 33 CFR 209.120(d)(11) and H. Rept. 91-113, p. 61-62.
[4]This regulation, which became effective 18 December 1968, revised the corps regulation of 7 December 1967, which had read (sec. 209.330(a)):
"The decision as to whether a permit will be issued will be predicated upon the effects of the permitted activities on the public interest including effects upon water quality, recreation, fish and wildlife, pollution, our natural resources, as well as the effects on navigation . . . ."

recently rejected the efforts of land developers to fill in a major part of Boca Ciega Bay, near St. Petersburg, Fla. See *Zabel v. Tabb*, 296 F. Supp. 764 (D.C.M.D.Fla., Tampa Div., Feb. 17, 1969), now on appeal to the U.S. Court of Appeals, Fifth Circuit, No. 27555.

The committee commends the Corps for recognizing its broader responsibilities to protect against unnecessary fills and other alteration of water bodies. However, the committee is concerned whether the Corps carries out in practice what its new regulations so properly state. There have been recent instances where there appeared to be a substantial gap between promise and performance. E.g., see the committee's report of March 24, 1969 (H.Rept. 91-113), entitled "The Permit for Landfill in Hunting Creek: A Debacle in Conservation." The committee therefore recommends as follows:

The Corps of Engineers should instruct its district engineers and other personnel involved in considering applications for fills, dredging, or other work in estuaries, rivers, and other bodies of navigable water to increase their emphasis on how the work will affect all aspects of the public interest, including not only navigation but also conservation of natural resources, fish and wildlife, air and water quality, esthetics, scenic view, historic sites, ecology, and other public interest aspects of the waterway.

The committee also believes that the Corps of Engineers must make an about-face in its handling of applications for new landfills, dredging, or other work in navigable water. The Corps has often routinely approved such applications unless the *opponents* of the permit clearly showed that substantial damage to the public interest will result. The committee believes that the Corps should place on the *applicant* the burden of proving that the filling, dredging, or other work is indisputably in accord with the public interest. The Corps should be sure that the environment will not be substantially harmed; or that there is no feasible and prudent alternative to such work and that all possible measures will be taken to minimize the resulting harm. In arriving at such judgment, the Corps should evaluate the relationship of the proposed work to the entire waterway and the total environment. Therefore the committee recommends as follows:

The Corps of Engineers should permit no further landfills or other work in the Nation's estuaries, rivers and other waterways except in those cases where the applicant affirmatively proves that the proposed work is in accord with the public interest, including the need to avoid the piecemeal destruction of these water areas.

II. THE "HARBOR LINE" PROCEDURES HERETOFORE FOLLOWED BY THE CORPS OF ENGINEERS DID NOT ADEQUATELY PROTECT AGAINST THE FILLING OF SUBSTANTIAL AREAS OF SUBMERGED LANDS SHOREWARD OF THE HARBOR LINES, AND VIOLATED THE FISH AND WILDLIFE COORDINATION ACT

Section 11 of the River and Harbor Act of 1899 (33 U.S.C. 404) authorizes the Secretary of the Army to establish "harbor lines," whenever he deems them "essential to the preservation and protection of harbors." The same section states that "beyond" such lines "no piers, wharves, bulkheads, or other works shall be extended or deposits made, except under such regulations" as he prescribes.

The Corps of Engineers and the Secretary of the Army have established three types of harbor lines:

(1) Pierhead lines, to mark the limits of open-pile work;
(2) Bulkhead lines, to mark the limits of solid fill;
(3) Pierhead-bulkhead lines, shoreward of which either open or solid construction is permissible.

The Corps' regulations state that the establishment of such harbor lines "implies consent to riparian owners to erect structures to the line without special authorization . . ." 33 CFR 209.150 (i).[5]

There are large areas of submerged lands which are shoreward of established bulkhead lines and pierhead-bulkhead lines. For example, the committee's hearings on San Francisco Bay disclosed that the area of such lands in the bay exceeds 19 square miles. Under the Corps' regulation the owners of these 19 square miles are apparently at liberty to fill them without seeking a permit from the Corps of Engineers.

The committee believes that the Corps' largely laissez-faire policy concerning landfills and construction on submerged lands and tidelands landward of harbor lines violated its statutory responsibility to protect all aspects of the public interest in those lands. Furthermore, the Corps' failure to consult with the Fish and Wildlife Service on all proposed work which modifies or affects those lands violated the intent of the Fish and Wildlife Coordination Act. That act, as amended in 1958, requires the Corps to review and consult with the Federal and State agencies having jurisdiction over wildlife resources "whenever the waters of any . . . body of water are . . . to be impounded, diverted . . . or . . . controlled or modified for any purpose whatever . . . " and makes no exception as to submerged lands shoreward of harbor lines. (16 U.S.C. 662(a)).

In areas of navigable waters where harbor lines do not exist the Corps has announced its intention to comply with the policies and procedures enunciated in this act and in the 1967 Memorandum of Understanding which the Secretary of the Army and the Secretary of the Interior entered into to better effectuate those policies.

Large sections of the submerged lands shoreward of established harbor lines have characteristics which, hydrologically and ecologically, are substantially similar to those of many areas of submerged lands where harbor lines do not

---

[5]However, the Corps' regulations also warn that the harbor lines do "not imply consent to operations of every kind landward of the line," and mention dredging work as an example of operations which "will ordinarily require the authorization of the Department to insure that operations are conducted under proper restrictions." 33 CFR 209.150(i)(1). In addition, the regulations direct the district engineers to "keep informed. . . of operations landward of harbor lines" and to require advance submission of plans where "proposed structures are to touch or closely approach the harbor line." *Ibid.* subparagraph (2).

exist. Hence, it seems incredible that the Corps has not followed the same procedures with respect to all submerged lands regardless of harbor lines.

The committee questions whether the Corps' practice of allowing anyone to fill or construct structures landward of harbor lines without a permit has been valid since 1890.

Congress first provided for the establishment of harbor lines under Federal authority in section 12 of the act of August 11, 1888 (25 Stat. 425),[6] which read as follows:

> Where it is made manifest to the Secretary of War that the establishment of harbor lines is essential to the preservation and protection of harbors, he may, and is, hereby, authorized to cause such lines to be established, beyond which no piers or wharves shall be extended or deposits made except under such regulations as may be prescribed from time to time by him.

The legislative history of the 1888 act shows that Congress enacted section 12 to overrule the Supreme Court's ruling in *Willamette Iron Bridge Co. v. Hatch,* 125 U.S. 1 (March 19, 1888), that existing Federal law did not prohibit obstructions to navigable waters within a single State. It was intended to provide Federal legal protection for the open harbors which the Supreme Court had held to be lacking. It did not grant any legal right to owners of land shoreward of a harbor line to fill or erect structures. It simply did not deal with the areas shoreward of a harbor line.

In 1890 Congress went further, and reversed the *Willamette Iron Bridge* rule completely, by providing, in section 10 of the River and Harbor Act of September 19, 1890 (26 Stat. 454),[7] as follows:

> That the creation of any obstruction, not *affirmatively* authorized by law, to the navigable capacity of any waters, in respect of which the United States has jurisdiction, is hereby prohibited. [Emphasis supplied].

Sections 6 and 7 of the 1890 act authorized the Secretary of War to issue permits for certain proposed structures, or for depositing refuse, in any navigable water, *irrespective of where a harbor line might be.* However, instead of repealing section 12 of the act of 1888, Congress reenacted it in section 12 of the 1890 statute, adding a criminal penalty.

Nine years later, Congress codified the Federal laws relating to navigable waters as sections 9 to 20 of the River and Harbor Act of 1899 (act of March 3, 1899, 30 Stat. 1151-1155). Despite their apparent incongruity, both the section of the 1890 law forbidding "any obstruction not affirmatively authorized" and the harbor line section (which does not give "affirmative" authorization for

---

[6] An earlier law (act of Aug. 5, 1886. c. 929, sec. 2,24 Stat. 329, 33 U.S.C. 407a) authorized the Secretary of War "in places where harbor lines have not been established" to establish lines limiting dump grounds for "debris of mines or stamp works" in navigable water. The "harbor lines" referred to in this statue were those established under State or local law. See Engvs. Peckham, 11 R.I. 210, 224 (1875).

[7] The legislative history of sec. 10 is reviewed in United States v. Republic Steel Corp., 362 U.S. 482 (1960), and United States v. Standard Oil Co., 32 U.S. 224 (1966).

obstructions shoreward of the harbor lines) were incorporated, with revisions, as section 10 and section 11 of the 1899 act, respectively. Both these sections are still in force. (33 U.S.C. 403, 404).

The 1899 act was infelicitously drafted and its ambiguities and overlapping language have caused much litigation and required much judicial creativeness.[8] The incongruity of the harbor lines language, insofar as concerns the validity of landfills or structures landward of the harbor lines, caused at least one court of appeals to say:[9]

> The argument is not without force that while this authorization to the Secretary of War is negative in form, yet when it is taken in connection with the power conferred on him to grant permission, under regulations prescribed by him, to erect structures beyond the harbor line, it implies affirmative authorization to the riparian owner to erect structures to the harbor line without special authorization.

Nevertheless, the court rejected the contention that such authorization gives the riparian owner a vested right to keep his fill or structures where he has placed them or to demand compensation from the Government if the harbor line is changed. The rule was stated by the Attorney General as follows (22 Ops. Atty. Gen. 501, 510, June 8, 1899):

> It is doubtless true that the establishment by the Government of a harbor line, within and out to which wharves, docks, et cetera, may be built by riparian owners, is an invitation and authority by the Government for the erection of such structures in aid of commerce. But, in view of this continuing power and duty of Congress to regulate this detail of commerce, it must be taken that such structures are erected in view of, and subject to the exercise of, this power at some future time. And, in view of the frequently changing conditions which require changes of harbor lines, it cannot be claimed that the establishment of such a line gives to anyone a vested right in its permanent continuance. Doubtless the existence of such structures under such circumstances and by such authority, their cost, size, character, and sufficiency, and the effect upon them of the establishment of a different line, are important factors to be considered and balanced against the needs of commerce for such change in determining whether such change shall be made.
>
> But this does not affect the power, and if, after looking at both sides, it is deemed that the needs of commerce require such change, I cannot doubt the power to make it, even though its exercise should affect injuriously riparian owners and their property.

It is therefore quite clear that the establishment of a harbor line does not constitute an abandonment of the Federal interest in the water area shoreward of the harbor line.

[8]See e.g., United States v. Republic Steel Co., 362 U.S. 482 (1960); United States v. Standard Oil Co., 384 U.S. 224 (1966): Wyandotte Transportation Co. v. United States, 389 U.S. 191 (1967). The 1899 act was drafted in the office of the Chief of Engineers (see H. Doc. 293, 54th Cong., 2d sess.).

[9]Garrison v. Greenleaf Johnson Lumber Co., 215 F. 576, 580 (CCA 4th, 1914), aff'd. 237 U.S. 251 (1915).

The committee does not believe it is necessary to explore fully the ambiguities of this "implied authorization," because the Corps has ample power, under section 11 of the 1899 act (33 U.S.C. sec. 404) to revise its harbor line regulations.

The Corps of Engineers has, over the past 80 years, established harbor lines primarily to identify existing and prospective needs of navigation and to aid the orderly development of port facilities. In most instances, particularly those where harbor lines were established many years ago, the Corps emphasized navigation and port uses, and gave little or no attention to such matters as fish and wildlife, air and water pollution control, esthetics, ecology, conservation of natural and scenic resources, recreation needs, and other matters of public interest. The increasing impairment and degradation of the environment which is affecting our country has brought an increasing awareness that the mounting pressure for fills and construction of structures in water areas landward of established harbor lines may result in pollution, adversely affect fish and wildlife, change the ecology of the waters, impair the esthetic values of the area, and otherwise result in harm to the public interest.

The harbor lines themselves serve the useful purpose of showing the outer limits beyond which no bulkheads or piers will be allowed and thus aid in the planning of future developments of waterfront property and in protecting the harbor. However, these purposes do not require continuation of the Corps' general practice of allowing all fills and construction shoreward of the harbor lines without a permit from the Corps. The Corps' witnesses testified as follows at the committee's San Francisco Bay hearings:

Mr. *Reuss.* I am still puzzled. While it is entirely proper for the Corps to grant a fill permit for dock or pier purposes shoreward of the harbor line, it would seem to me the Corps is not doing its ecological job if it does not even ask that people come in and ask for a permit for housing, a non-water-borne industry, a junkyard, or whatever else somebody wants to build.

Colonel Boerger. Yes, sir; that is an inconsistency. (Hearings, May 15, 1969, p. 119.)

• • •

Mr. *Reuss.* . . . Referring just to the area shoreward of the harbor lines, both in the San Francisco Bay and everywhere else within the jurisdiction of the Corps of Engineers, why doesn't the Corps immediately abandon its present self-imposed rule that it will have nothing to say about what happens shoreward . . . of a harbor line . . . ? Why cut yourself off from a complete review simply by reason of the fact that in certain areas you happen to have erected harbor lines in the past for good and sufficent reasons?

Colonel Meanor. I follow your line of questioning. I think it is something that we have got to start doing. . . . (Hearings, May 15, 1969, pp. 131-132.)

• • •

Mr. *Reuss.* At any rate there is a tremendous area back of the bulkhead lines, much of it of great ecological significance. And it seems to me vital that the Corps in San Francisco Bay and nationally assert its power to weigh and consider and protect the public environmental interest and not permit fills—whether they are back of a harbor line or not back of a harbor line—which are going to ruin the environment. Would you not agree? *General Glasgow.* I would, Mr. Chairman. (Hearings, August 20, 1969, p. 63.)

The committee does not believe it is necessary for the Corps to abolish all existing harbor lines, or even to go through a procedure (which would necessarily be lengthy and time-consuming) of reexamining and, after public hearing, revalidating, the existing harbor lines. Instead, the committee believes, and recommends as follows:

The Corps of Engineers should revise its regulations to (a) make harbor lines merely guidelines defining the offshore limits of bulkheads, fills, piers, and other structures, and (b) require anyone planning to do any work (including filling) shoreward of a harbor line to apply and obtain a permit for such work, subject to such conditions as the Corps deems necessary to protect the public interest. In reviewing all applications for such permits, the Corps should comply with the same interdepartmental review and consultation procedures as are used in considering applications for permits for similar work in waters where harbor lines are not established.

The committee understands that the Corps is now proceeding to amend its regulations in conformity with the foregoing recommendation.

III. THE CORPS OF ENGINEERS HOLDS PUBLIC HEARINGS, WHENEVER THERE IS SUFFICIENT PUBLIC INTEREST, ON APPLICATIONS FOR PERMITS TO FILL, DREDGE OR CONSTRUCT STRUCTURES IN NAVIGABLE WATERS, BUT AVOIDS SUCH HEARINGS IN CONNECTION WITH APPLICATIONS TO ESTABLISH OR MODIFY HARBOR LINES WHICH, UNDER PRESENT CORPS PRACTICE, CAN RESULT IN LANDFILLS OR CONSTRUCTION OF STRUCTURES LANDWARD OF THE HARBOR LINE WITHOUT FURTHER CORPS REVIEW

The Corps has an eminently sound policy concerning public hearings on applications for landfills, dredging, and other proposed work in navigable waters. Its regulations state (33 CFR 209.120(g)(1)):

> (g) *Public hearing.* (1) It is the policy of the Chief of Engineers to conduct his civil works program in an atmosphere of public understanding, trust, and mutual cooperation, and in a manner responsive to public needs and desires. To this end, public hearings are helpful and will be held whenever there appears to be sufficient public interest to justify such action. *In case of doubt, a public hearing will be held.* (Emphasis added.)

Oddly, however, the Corps' regulations enunciate a contrary position with respect to the establishment of harbor lines. Although subparagraph (1) of the Corps' regulation (33 CFR 209.150(e)) requires that public notices of applications for establishment or modification of harbor lines must be issued "to all known interested parties," subparagraph (2) specifies as follows:

Public hearings in connection with harbor lines *will be kept to a minimum and will be the exception rather than the rule.* A hearing will be held at the discretion of the district engineer only if he deems a public hearing essential, as in a known controversial case or when response from the public notice indicates that a public hearing should be held. (Emphasis added.)

If a public hearing is warranted in cases where there is demonstrated public interest over the issuance of dredging or fill permits (and the committee believes it is), then it is equally important to hold such hearings when harbor lines are established or changed. The importance of such a hearing is especially evident in view of the fact that, as the Corps regulation expressly notes, "the establishment of a bulkhead line has been frequently followed by solid filling to the limit and by requests thereafter from riparian owners to push the limit farther toward the channel" (33 CFR, sec. 209.150(a)(5)).

The impact of the Corps' grudging policy concerning public hearings when harbor line changes are involved is illustrated by the Corps' treatment of protests against the proposal by the Port of Oakland in 1965 that the Corps modify the bulkhead lines on the east side of the San Francisco Bay. Such modification would have enabled the Port of Oakland to fill 140 acres of bay to expand its Seventh Street Marine Terminal.

The California State Department of Fish and Game and over 80 other organizations and persons filed protests against the proposed harbor line revision and fill. The Fish and Game Department particularly urged that the fill would adversely affect fish and wildlife resources and that further harbor development should be consistent with a comprehensive bay plan. Later, the Governor designated the California Resources Agency as his spokesman on this harbor line application. The director of that agency informed the Corps "that it was in the best interest of the public to construct the terminal" on the filled area. The district engineer's letter to the division engineer on this project noted that all differences were resolved, but did not mention the protest of the State Fish and Game Department.

In response to a question by Chairman Reuss in his letter of July 31, 1969, the Corps states that "because of the resolution of these differences, it was not considered necessary that a public hearing be held." (Hearings, Part 2, August 20, 1969, pp. 7-8.) The Corps apparently simply ignored the other 80 protests which, the Corps said, were "of a general nature . . . primarily from individuals interested in conservation." General Glasgow testified as follows at the subcommittee's San Francisco Bay hearings (hearings, *ibid,* pp. 73-74):

General Glasgow. These letters, as I stated, were primarily from persons who had interests in conservation, many of whom were opposed to fill of the Bay for any reason or any purpose. These views, we felt, were also considered in the views of the State of California Resources Agency, and you will recall I stated that we also met with them, that the director of the Resources Agency had been designated by the Governor of the State to represent fish and game as well as other interests of the State. We felt that when the State of California Resources Agency then withdrew their objection that we then were aware of the opposition of people. We were aware of the basis of their opposition—

Mr. Moss. How?

General Glasgow.—and that the State had determined, in balancing these concerns against economic advantages to the State and so forth, and they told us to proceed in favor of the fill.

Mr. Moss. They couldn't tell you to proceed, they could merely indicate that their objections—

General Glasgow. They withdrew their objections, that is right, sir.

● ● ●

Mr. Moss. The 80 letters you regarded as being so biased as not to constitute a basis for public hearing?

General Glasgow. No, sir. No, sir.

Mr. Moss. Well, you have stated there was a clear—it was quite clear, that the majority were opposed to a fill for any reason. Therefore, you were inferring that there was a bias.

General Glasgow. I would hesitate to use that term. I was saying that in our judgment the concerns of those 80 letters or protests were considered and reflected in the concern of the California State Department of Fish and Game, and that in balance it was the State's judgment that they would withdraw their objection, and we accepted that withdrawal of the objection.

The Corps thus accepted the State Resources Agency's judgment concerning the needs of the public as superior to that of 80 individual protestors who, the Corps apparently believed, were "opposed to fill of the bay for any reason or any purpose. The Corps gave them no opportunity to air their views in a public hearing. As Congressman Moss stated during the hearing (*ibid,* p. 75):

. . . I think there is a matter of transcendant importance here; namely, the public right of protest, and the right to have the protest heard with reasonable impartiality.

The absence of a public hearing was compounded by the district engineer's failure to include in his letter to the division engineer any mention of the protest by the California State Fish and Game Department, thus further submerging the fish and wildlife interest when the harbor line proposal was reviewed by the district engineer's superiors.

The committee believes that the protection of the Nation's estuaries and other bodies of water requires that public hearings be held and encouraged whenever there is sufficient public interest in the establishment or modification of a harbor line, and that any doubts as to the sufficiency of the public interest in the harbor line procedure be resolved in favor of holding a hearing. The committee also believes that the Corps officers who must decide these problems should have the full record before them. The committee therefore recommends as follows:

a. **The Corps of Engineers should promptly revise its regulations to require and encourage public hearings on proposals to establish or modify harbor lines whenever there is sufficient public interest in such proposals.**

b. **The Corps of Engineers should make sure that the record on each application for a permit, harbor line change, or other action contains all recommendations and objections received by the Corps.**

This recommendation is in accord with the President's recent Executive Order (No. 11514, March 5, 1970, 35 F.R. 4247) directing all Federal agencies to develop procedures for public hearings on activities affecting the quality of the environment. Sec. 2(b).

IV. AT THE COMMITTEE'S REQUEST, THE CORPS OF ENGINEERS IS AMENDING ITS REGULATIONS TO REQUIRE THAT APPLICANTS FOR PERMITS TO CONSTRUCT SEWER OUTFALLS INTO NAVIGABLE WATERWAYS MUST FURNISH INFORMATION ADEQUATELY DESCRIBING THE EFFLUENT TO BE DISCHARGED THROUGH THE OUTFALL

The Corp's present regulations state that every application for a permit to fill, dredge, or do other work in navigable waters will be evaluated in light of "all relevant factors, including the effect of the proposed work on navigation, fish and wildlife, conservation, pollution, esthetics, ecology, and the general public interest." 33 CFR 209.120(d). However, applicants for permits to construct sewer outfalls "which may affect the navigable capacity of a waterway" have not been required to furnish adequate information concerning the effluent to be discharged from the proposed sewer outfall. The present regulation has required only information on the amount of sand or sediments to be discharged. 33 CFR 209.130(b)(18)(iii). Such information would enable the Corps to evaluate the possibility that such sediments will silt up the navigation channels. But it would hardly enable the Corps to fully evaluate the potential effect of the sewer outfall on fish and wildlife, pollution, esthetics, ecology, and other public interests which may be affected by nonsedimentary effluents (such as chemicals) discharged through the proposed outfall, or whether vessels would be corroded, or seamen's lives endangered, by corrosive or inflammable chemicals.

In the fall of 1969, Dr. Joel W. Hedgpeth, director of the Marine Science Center, Oregon State University, Newport, Oreg., complained to this committee that he had been unable to obtain information concerning the chemical wastes that would be discharged from a proposed sewer outfall which Virginia Chemicals, Inc., desired to construct in the Columbia River. Chairman Reuss thereupon wrote to the Chief of Engineers pointing out that the Corps' regulation governing the information required from the applicant was inadequate for both the Corps' evaluation of the application and the public's right to know the composition and amount of wastes to be discharged into the public waterways through the proposed sewer outfall. On December 17, 1969, the Acting Chief of Engineers acknowledged that "an applicant is not specifically required to identify the effluent that will be discharged," and stated that the Corps' regulation would be revised to "eliminate this imprecision" and "particularize the requirement."

The committee commends the Corps on this progressive step to assure that it will more effectively carry out its responsibilities to evaluate the potential effects which new sewer outfalls may have on water quality, fish and wildlife, esthetics, ecology, and other public interests in the waterways receiving waste discharges from such outfalls. The information which will be required under the revised regulation will also enable Federal and State water pollution control agencies to do their job more effectively.[10]

The committee recommends as follows:

Before granting a permit to construct a sewer outfall, the Corps of Engineers should:

(a) require the applicant to furnish full information concerning the nature, composition, amount and degree of treatment of the wastes which will be discharged from the outfall, and to demonstrate that such discharges will not adversely affect the quality of the receiving waters;

(b) consult with the appropriate Interior Department agencies (Federal Water Pollution Control Administration, Fish and Wildlife Service, etc.) as to whether, and under what conditions, the permit should be granted; and

(c) specify in the permit that the permittee shall furnish, upon the Government's request, full information from time to time concerning the wastes discharged through the outfall and comply with all requirements concerning protection of the quality of the receiving waters.

V. THE CORPS OF ENGINEERS CAN SUBSTANTIALLY HELP TO PREVENT POLLUTION OF OUR NATION'S WATERS BY VIGOROUSLY ENFORCING THE REFUSE ACT OF 1899 WHICH PROHIBITS DISCHARGE OF REFUSE INTO NAVIGABLE WATERS, AND DEPOSIT OF POLLUTING MATERIALS ON THEIR BANKS

Section 13 of the River and Harbor Act of 1899 (30 Stat. 1151, 33 U.S.C. 407) although enacted nearly three quarters of a century ago, constitutes a potentially powerful, but only sporadically used, weapon for combatting the pollution of our Nation's navigable waters. Section 13, commonly called the Refuse Act, states:

It shall not be lawful to throw, discharge, or deposit, or cause, suffer, or procure to be thrown, discharged, or deposited either from or out of any

---

[10]See this committee's report entitled* The critical need for a national inventory of industrial wastes (water pollution control and abatement).* H. Rept. 1579, 90th Cong., June 24, 1968.

ship, barge, or other floating craft of any kind, or from the shore, wharf, manufacturing establishment, or mill of any kind, any refuse matter of any kind or description whatever other than that flowing from streets and sewers and passing therefrom in a liquid state, into any navigable water of the United States, or into any tributary of any navigable water from which the same shall float or be washed into such navigable water; and it shall not be lawful to deposit, or cause, suffer, or procure to be deposited material of any kind in any place on the bank of any navigable water, or on the bank of any tributary of any navigable water, where the same shall be liable to be washed into such navigable water . . . whereby navigation shall or may be impeded or obstructed . . . [11]

Judicial interpretation over the years has enhanced the usefulness of the Refuse Act in controlling and abating pollution of our waterways. Thus, the courts have ruled that the offense of discharging refuse into navigable waters is not limited, as is the offense of depositing material on the banks of navigable waters, by the language "whereby navigation shall or may be impeded or obstructed."[12]

The prohibition against depositing "refuse" in navigable waters has been held to apply to industrial fuels or chemicals which were commercially valuable at the time they were deposited into the navigable waters.[13] The Supreme Court in 1966 stated flatly:

The word "refuse" includes *all foreign substances and pollutants* apart from those "flowing from streets and sewers and passing therefrom in a liquid state" into the watercourse. (Emphasis supplied.)[14]

Furthermore, the latter exception has been narrowly construed. Thus, the Supreme Court, in holding that the Refuse Act was violated by a discharge through non-municipal sewers of industrial wastes containing suspended solids which settled into navigable waters, said: "Refuse flowing from 'sewers' in a 'liquid state' mean to us 'sewage'."[15]

Moreover, the Refuse Act has been used successfully against companies whose

[11] A proviso of Section 13 authorizes the Secretary of the Army to permit deposit of material in navigable water "under conditions to be prescribed by him," if the Chief of Engineers advises that "anchorage and navigation will not be injured," and specifies that "whenever any permit is so granted the conditions thereof shall be strictly complied with, and any violation thereof shall be unlawful."

[12] *The La Merced* (United States v. Alaska Southern Packing Co.), 84 F. 2d 444, 445-446 (C.A.9, 1936).

[13] *The La Merced, supra* (oil discharged from a ship); United States v. Ballard Oil Co. of Hartford, Inc., 195 F. 2d 369, 372 (CA2, 1952) (oil overflowing from an on-shore tank into the Connecticut River); United States v. Standard Oil Co., 384 U.S. 224 (1966) (accidental discharge of commercially valuable 100 octane aviation gasoline into the St. John's River, Florida). Similar rulings were made in *The Albania*, 32 F. 2d 727 (D.C. S.D. N.Y. 1928) and *The Columbo*, 42 F. 2d 211 (C.A. 2, 1930), with respect to the discharge of oil, under the Act of June 29, 1888 (25 Stat. 209, 33 U.S. code 441), which prohibits discharge of "refuse" into New York Harbor.

[14] United States v. Standard Oil Co., 384 U.S. 224, 230 (1966).

[15] United States v. Republic Steel Corp., 362 U.S. 482, 490 (1960).

employees dumped garbage from a ship,[16] or allowed oil to spill from shore storage tanks and flow "indirectly" (i.e. over land by force of gravity) into navigable waters.[17] The violation occurs even though the discharge was not done intentionally, negligently or knowingly.[18]

It is apparent that the Refuse Act, which prohibits the discharge or deposit into navigable waters of "all foreign substances and pollutants," as the Supreme Court put it—including oil but not sewage—is a broad charter of authority and a powerful legal tool for preventing the pollution of all navigable waters.

Its present usefulness is not reduced by more recent water pollution control legislation. The Federal Water Pollution Control Act specifically states (33 U.S. Code sec. 466k) that it "shall not be construed as (1) superseding or limiting the functions, under any other law—of any other officer or agency of the United States, relating to water pollution, or (2) affecting or impairing the provisions of . . . sections 13 through 17 of the" River and Harbor Act of 1899, as amended (i.e., the Refuse Act.[19]

The Federal Water Pollution Control Act authorizes the Secretary of the Interior to promote the control and abatement of water pollution by a wide variety of methods, including encouraging interstate cooperation and uniform State laws; sponsoring research, training and development of means of pollution control; studying estuarine areas; granting funds to the States for research and development, and administration of pollution control programs, and to municipalities for construction of water treatment facilities; approving and enforcing State water quality standards for interstate waters; holding conferences and public hearings for abating pollution; and, in certain circumstances, requesting the Attorney General to bring injunction suits against polluters.

However, the Federal Water Pollution Control Act contains various limitations on its scope and enforcement powers. For example, it requires water quality standards only for interstate waters. Furthermore, it provides that discharges of wastes into interstate waters which reduce their quality below established water quality standards are subject to abatement only after notice and a waiting period of at least 180 days. The abatement proceedings may be instituted only upon the Governor's consent unless the pollution "is endangering the health or welfare of persons in a State other than that in which the discharge or discharges . . . originate". Moreover, the court in such abatement proceedings need not confine itself to examing the issues of law and facts, but is authorized to give "due consideration to the practicability and to the physical and

---

[16]*The President Coolidge* (Dollar S.S. Co. v. United States), 101 F. 2d 638 (C.A. 9, 1939) (garbage dumped from ship by employee in disregard of company's orders and without knowledge of ship's officers).

[17]United States v. Ballard Oil Co. of Hartford, Inc.., 195 F. 2d 369 (C.A. 2, 1952); United States v. Esso Standard Oil Company of Puerto Rico, 375 F. 2d 621 (C.A. 3, 1967).

[18] United States v. Interlake Steel Corp., 297 F. Supp. 912 (D.C., N.D. Ill. E.D. 1969); *The President Coolidge, supra.*

[19]United States v. Interlake Steel Corp., *supra* at 916.

economic feasibility of complying" with the established water quality standards as well as reviewing the standards themselves.[20]

Similarly, the Oil Pollution Act, 1924,[21] specifically provided that it "shall be in addition to other laws for the preservation and protection of navigable waters of the United States and shall not be construed as repealing, modifying, or in any manner affecting the provisions of such laws." (33 U.S. Code, Supp. IV, sec. 437). This Act is a narrowly drawn criminal statute prohibiting, with certain exceptions.[22] only the "grossly negligent, or willful" discharge of oil from a boat or vessel into navigable waters. Violators of the statute must clean up their oil spills or pay the United States for the cost of doing so. They are subject to a $2500 fine or one year's imprisonment or both, and the master of the vessel may have his license revoked. The Corps of Engineers, Coast Guard, and Bureau of Customs assist the Department of the Interior in enforcing the statute.

The proposed Water Quality Improvement Act now being considered by House-Senate conferees (H.R. 4148) will enact new sections in the Federal Water Pollution Control Act to govern oil pollution and hazardous substances and repeal the Oil Pollution Act, 1924.

The Refuse Act, which prohibits discharges of all foreign material except sewage into any navigable water of the United States,[23] is not subject to many of the limitations present in the Federal Water Pollution Control Act and the Oil Pollution Act, 1924, and therefore provides protection against pollution of waterways in many circumstances where the other acts are more circumscribed.

The committee therefore recommends:

The Corps of Engineers should vigorously enforce the Refuse Act of 1899 which prohibits discharge of refuse into navigable waters and deposit of polluting materials on their banks.

There are several methods which the Corps can utilize under the Refuse Act to prevent pollution of waterways:

[20] 33 U.S.C. 466g (c) (5) and (g). On 10 February 1970, the President proposed legislation that would partially modify this enforcement procedure. See H.R. 15872, 91st Cong.

[21] Oil Pollution Act, 1924, as amended by the Clean Water Restoration Act of November 3, 1966 (Sec. 211, Public Law 89-753; U.S. Code, Supp. IV, sec. 431 et seq.).

[22] See 33 U.S. Code, sec. 433.

[23] The Courts have held that navigable waters include waterways which either in their natural or improved condition are used, or can be used, for floating light boats or logs, even though the waterway may be obstructed by falls, rapids, sand bars, currents, etc., and even though the waterway has not been used for navigation for many years. United States v. Appalachian Electric Power Co., 311 U.S. 377, 407-410, 416 (1940); Wisconsin Public Service Corp. v. Federal Power Commission, 147 F. 2d 743 (CA 7; 1945), cert den. 325 U.S. 880; Wisconsin v. Federal Power Commission, 214 F. 2d 334 (CA 7, 1954), cert. den. 348 U.S. 883 (1954); Namekagon Hydro Co. v. Federal Power Commission, 216 F.2d 509 (CA 7, 1954); Puente de Reynosa, S.A. v. City of McAllen, 357 F.2d 43, 50-51 (CA 5, 1966); Rochester Gas and Electric Corp. v. Federal Power Commission, 344 F.2d 594 (CA 2, 1965).

1. Violations of the Refuse Act are subject to criminal prosecution and penalties of fine not exceeding $2,500 nor less than $500, or imprisonment for not less than 30 days nor more than one year, or both such fine and imprisonment (33 U.S. Code 411). Officers and employees of the Corps of Engineers are authorized "to arrest and take into custody" and request prosecution of, and it is "the duty of United States attorneys to vigorously prosecute all offenders" of the Refuse Act. (33 U.S. Code 413).[24] The law further specifies that "one-half of said fine" shall be "paid to the person or persons giving information which shall lead to conviction." This provision buttresses the Corps' efficacy in carrying out the Refuse Act in two ways:

(a) The informer payment provides a monetary incentive to citizens to furnish information to the Corps concerning violations of the Refuse Act.

(b) The Supreme Court has ruled that where a statute provides for a reward to the informer, the statute authorizes him, if the Government has not previously instituted a prosecution against the violator,[25] to institute his own suit in the name of the United States (a *qui tam* action) to collect his moiety of the penalty.[26] Such *qui tam* statutes, vesting in an informer the right to recover a moiety of a penalty for a violation in which he otherwise would have no financial interest, "have been in existence for hundreds of years in England, and in this country ever since the foundation of our Government."[27] By making the violator subject to action by private persons stimulated by the hope of a reward, such provisions help to insure against laxity by public officials in enforcing statutes effectuating important public policies.

2. The Supreme Court has ruled that the Federal Government may obtain injunctions requiring a polluter, who has discharged into a navigable waterway a foreign substance or pollutant prohibited by the Refuse Act, to cease future discharges and to remove the polluting substance already discharged.[28] The Federal Government has rarely sought injunctions as a method of controlling or abating pollution. However, the committee believes that the Nation's fight against water pollution would be greatly strengthened and advanced by more frequent use of this authority under the Refuse Act which is not subject to the limitations of the Federal Water Pollution Control Act.

[24]The Coast Guard assists the Corps in detecting and reporting violations of the Refuse Act. The duty of the U.S. Attorney to prosecute offenders under the Refuse Act is not dependent on whether the prosecution is requested by the Corps of Engineers, or by any other Federal agency or employee, or whether the alleged offense was contrary to any water quality standards set by a State agency under the Water Quality Act of 1965. United States v. Interlake Steel Corp., *supra*, pp; 914, 916.

[25]Francis v. United States, 72 U.S. (5 Wall.) 338 (1866).

[26]Adams, qui tam v. Woods, 6 U.S. (1 Cranch) 336 (1805); United States ex rel. Marcus v. Hess, 317 U.S. 537. 541 (1943).

[27]Marrin v. Trout, 199 U.S. 212, 225 (1905); United States ex rel. Marcus v. Hess, 317 U.S. 537, 541 (1943).

[28]United States v. Republic Steel Corp., 362 U.S. 482 (1960); Wyandotte Transportation Co. v. United States, 389 U.S. 191, 203-204, ftnt. 15 (1967).

The committee therefore recommends:

Both the Corps of Engineers and the Federal Water Pollution Control Administration should request the Attorney General to institute injunction suits against all persons whose discharges or deposits (except minor ones) violate the Refuse Act and are not promptly cleaned up or stopped by the polluter.

3. In addition to using criminal sanctions to punish for past discharges, and injunctions to preclude future discharges and remove pollutants already discharged, the Government can protect the Nation's navigable waters from pollution or degradation by calling upon the polluter to clean up the discharge voluntarily. If the polluter does not do so, the Government can itself do the clean up work and then, if the polluter's discharges were willful or negligent, bill the polluter for the Government's costs in doing such clean up work.[29] The use of this remedy, of course, entails expenditure of Government funds to perform clean up work. The Corps may sometimes find it difficult to perform such clean up promptly, because of limited appropriations. Furthermore, any reimbursement received by the Government would be covered into the Treasury instead of replenishing the funds available to the Corps. However, in many cases clean up by the Corps may be the only way to achieve prompt removal of pollutants discharged into a waterway or deposited on its banks.[30]

The committee therefore recommends:

The Corps of Engineers should proceed to increase its capability, including seeking the necessary contingency funds, to enable it to promptly remove or clean up pollutional discharges and deposits and to seek reimbursement of the costs thereof from persons who willfully or negligently made or caused such discharges or deposits.

---

[29]Wyandotte Transportation Co. v. United States, 389 U.S. 191 (1967).

[30]The proposed Water Quality Improvement Act (H.R. 4148), now being considered by House-Senate conferees, would establish a revolving fund to finance the Government's clean up and removal of oil and hazardous substances from waterways, and would require polluters to reimburse the fund.

Epilog

POLLUTED
WATER
'NO SWIMMING'
OR WADING

NATIONAL PARK SERVICE

A Look into Future Environmental Monitoring

We have seen from the previous readings that our environment is stressed in many ways due to a multitude of causes and that the nature and severity of the stress is measured and evaluated by many different parameters. The reader will recognize, however, that despite the wealth of information available to him in the literature, including the citations in this book, it is still not possible either to accurately delineate all areas of water pollution or to quantify the impact of many environmental stresses.

The quantative data presented herein, as we have seen, consist almost entirely of point source analyses of physical, chemical, or biological parameters as represented, for example, by an analysis of a water sample from a single well. Further, hydrology, being dynamic by definition, is concerned with changes in water quantity and quality characteristics. However, in order to measure the continuing impact of an environmental stress on water quality, it is necessary to observe the distribution of quality characteristics spatially (areally) and also to observe changes in these characteristics with time. Traditional water-quality monitoring techniques commonly do include repetitive or, in a few instances, continuous measurement of selected quality parameters. On the other hand, the problem in observing quality distribution spatially has not been solved except in certain special instances involving physical characteristics such as oil spills, color changes, or floating debris. However, an emerging remote-sensing technology, combined with modern computerized data processing, appears to offer potential

Photo opposite from Ryan Photographic Service, Inc. Glenn Dale, Md.

for monitoring environmental conditions on a spatial, continuous, and real-time basis.

Remote-sensing technology deals, among other things, with the measurement of reflected or emitted radiation from water in the environment. The most notable application of the new technology has been in the detection and measurement of thermal anomalies involving the infrared portion of the electromagnetic spectrum. It has been possible, for example, to analyze the imagery collected over part of the Lake Ontario shoreline in New York to detect and delineate thermal plumes from power plants and discharging streams.

It must be kept in mind that the emitted thermal radiation is not the same as the kinetic thermal energy as measured by a thermometer inserted into the water mass. In addition, the emitted energy is observed spatially while the kinetic energy is measured at a point. However, the point data of the type traditionally measured serve as so-called ground truth in interpreting the remotely sensed data.

To date, it has been possible to relate remotely sensed data spatially only to the point measurements of selected physical parameters such as temperature, as described above, color, and turbidity. One can also infer the distribution of certain chemical quality characteristics from the physical data; for example, the movement of the tide in an estuary where the discharging fresh river water and the brackish or saline water of the estuary are of different temperatures. In this instance, we are using the distribution of radiant energy to determine the surficial distribution of chloride, the parameter most indicative of salinity. It would be more desirable to remotely measure directly chloride or any other dissolved chemical constituent. Unfortunately, such a capability does not exist at the present time. For example, without ground data to relate to the remotely sensed data, it is impossible to distinguish from either reflected or emitted radiation any differences in the chloride content between fresh river water, brackish estuary water, sea water, or brine discharging from a desalinization plant.

Research currently underway by the U.S. Geological Survey and the National Aeronautics and Space Administration is aimed at the ultimate capability of measuring such quality parameters remotely. This research involves analysis of spectra reflected from solar radiation at various wavelengths. More promising is research in laser spectroscopy to detect and measure chemical properties synoptically.

However, it is not always necessary to measure the traditional quality parameters directly to assess the effect of a specific stress on the environment. The emerging remote-sensing technology in many instances provides the capability of observing directly the effects of such stress. For example, polluted water commonly has detrimental effects on the ecosystem. In certain instances these effects can be measured by the changes in reflectance with time or by noting the differences between stressed and unstressed species.

The future will see remote sensors of various types in use on the ground, in aircraft, and in spacecraft, observing water areas and the surface features that come in contact with water. These sensors will directly monitor environmental conditions in both space and time. The water management implications are profound. Tomorrow brings promise of further exotic and sophisticated instrumentation and techniques that will permit Twenty-first century monitoring of twentieth century problems that we are presently trying to evaluate and solve with nineteenth century techniques and laws. Hope for tomorrow lies in educating our citizenry to the point where they request and demand updated laws—man-made laws that are able to cope with the man-made advances of instrumentation and technology.